MANAGING HUMAN RESOURCES IN THE FIELD OF NUCLEAR ENERGY

The following States are Members of the International Atomic Energy Agency:

AFGHANISTAN
ALBANIA
ALGERIA
ANGOLA
ANTIGUA AND BARBUDA
ARGENTINA
ARMENIA
AUSTRALIA
AUSTRIA
AZERBAIJAN
BAHAMAS
BAHRAIN
BANGLADESH
BARBADOS
BELARUS
BELGIUM
BELIZE
BENIN
BOLIVIA, PLURINATIONAL STATE OF
BOSNIA AND HERZEGOVINA
BOTSWANA
BRAZIL
BRUNEI DARUSSALAM
BULGARIA
BURKINA FASO
BURUNDI
CAMBODIA
CAMEROON
CANADA
CENTRAL AFRICAN REPUBLIC
CHAD
CHILE
CHINA
COLOMBIA
COMOROS
CONGO
COSTA RICA
CÔTE D'IVOIRE
CROATIA
CUBA
CYPRUS
CZECH REPUBLIC
DEMOCRATIC REPUBLIC OF THE CONGO
DENMARK
DJIBOUTI
DOMINICA
DOMINICAN REPUBLIC
ECUADOR
EGYPT
EL SALVADOR
ERITREA
ESTONIA
ESWATINI
ETHIOPIA
FIJI
FINLAND
FRANCE
GABON
GEORGIA
GERMANY
GHANA
GREECE
GRENADA
GUATEMALA
GUYANA
HAITI
HOLY SEE
HONDURAS
HUNGARY
ICELAND
INDIA
INDONESIA
IRAN, ISLAMIC REPUBLIC OF
IRAQ
IRELAND
ISRAEL
ITALY
JAMAICA
JAPAN
JORDAN
KAZAKHSTAN
KENYA
KOREA, REPUBLIC OF
KUWAIT
KYRGYZSTAN
LAO PEOPLE'S DEMOCRATIC REPUBLIC
LATVIA
LEBANON
LESOTHO
LIBERIA
LIBYA
LIECHTENSTEIN
LITHUANIA
LUXEMBOURG
MADAGASCAR
MALAWI
MALAYSIA
MALI
MALTA
MARSHALL ISLANDS
MAURITANIA
MAURITIUS
MEXICO
MONACO
MONGOLIA
MONTENEGRO
MOROCCO
MOZAMBIQUE
MYANMAR
NAMIBIA
NEPAL
NETHERLANDS
NEW ZEALAND
NICARAGUA
NIGER
NIGERIA
NORTH MACEDONIA
NORWAY
OMAN
PAKISTAN
PALAU
PANAMA
PAPUA NEW GUINEA
PARAGUAY
PERU
PHILIPPINES
POLAND
PORTUGAL
QATAR
REPUBLIC OF MOLDOVA
ROMANIA
RUSSIAN FEDERATION
RWANDA
SAINT KITTS AND NEVIS
SAINT LUCIA
SAINT VINCENT AND THE GRENADINES
SAMOA
SAN MARINO
SAUDI ARABIA
SENEGAL
SERBIA
SEYCHELLES
SIERRA LEONE
SINGAPORE
SLOVAKIA
SLOVENIA
SOUTH AFRICA
SPAIN
SRI LANKA
SUDAN
SWEDEN
SWITZERLAND
SYRIAN ARAB REPUBLIC
TAJIKISTAN
THAILAND
TOGO
TONGA
TRINIDAD AND TOBAGO
TUNISIA
TÜRKİYE
TURKMENISTAN
UGANDA
UKRAINE
UNITED ARAB EMIRATES
UNITED KINGDOM OF GREAT BRITAIN AND NORTHERN IRELAND
UNITED REPUBLIC OF TANZANIA
UNITED STATES OF AMERICA
URUGUAY
UZBEKISTAN
VANUATU
VENEZUELA, BOLIVARIAN REPUBLIC OF
VIET NAM
YEMEN
ZAMBIA
ZIMBABWE

The Agency's Statute was approved on 23 October 1956 by the Conference on the Statute of the IAEA held at United Nations Headquarters, New York; it entered into force on 29 July 1957. The Headquarters of the Agency are situated in Vienna. Its principal objective is "to accelerate and enlarge the contribution of atomic energy to peace, health and prosperity throughout the world".

IAEA NUCLEAR ENERGY SERIES No. NG-G-2.1 (Rev. 1)

MANAGING HUMAN RESOURCES IN THE FIELD OF NUCLEAR ENERGY

INTERNATIONAL ATOMIC ENERGY AGENCY
VIENNA, 2023

COPYRIGHT NOTICE

Marketing and Sales Unit, Publishing Section
International Atomic Energy Agency
Vienna International Centre
PO Box 100
1400 Vienna, Austria
fax: +43 1 26007 22529
tel.: +43 1 2600 22417
email: sales.publications@iaea.org
www.iaea.org/publications

Printed by the IAEA in Austria
January 2023
STI/PUB/1958

IAEA Library Cataloguing in Publication Data

Names: International Atomic Energy Agency.
Title: Managing human resources in the field of nuclear energy / International Atomic Energy Agency.
Description: Vienna : International Atomic Energy Agency, 2023. | Series: IAEA nuclear energy series, ISSN 1995–7807 ; no. NG-G-2.1 (Rev. 1) | Includes bibliographical references.
Identifiers: IAEAL 22-01552 | ISBN 978–92–0–126121–2 (paperback : alk. paper) | ISBN 978–92–0–126221–9 (pdf) | ISBN 978–92–0–126321–6 (epub)
Subjects: LCSH: Nuclear industry — Human capital. | Nuclear industry — Employees. | Nuclear power plants — Employees. | Nuclear energy.
Classification: UDC 621.039:005.96 | STI/PUB/1958

FOREWORD

The IAEA's statutory role is to "seek to accelerate and enlarge the contribution of atomic energy to peace, health and prosperity throughout the world". Among other functions, the IAEA is authorized to "foster the exchange of scientific and technical information on peaceful uses of atomic energy". One way this is achieved is through a range of technical publications including the IAEA Nuclear Energy Series.

The IAEA Nuclear Energy Series comprises publications designed to further the use of nuclear technologies in support of sustainable development, to advance nuclear science and technology, catalyse innovation and build capacity to support the existing and expanded use of nuclear power and nuclear science applications. The publications include information covering all policy, technological and management aspects of the definition and implementation of activities involving the peaceful use of nuclear technology. While the guidance provided in IAEA Nuclear Energy Series publications does not constitute Member States' consensus, it has undergone internal peer review and been made available to Member States for comment prior to publication.

The IAEA safety standards establish fundamental principles, requirements and recommendations to ensure nuclear safety and serve as a global reference for protecting people and the environment from harmful effects of ionizing radiation.

When IAEA Nuclear Energy Series publications address safety, it is ensured that the IAEA safety standards are referred to as the current boundary conditions for the application of nuclear technology.

The unique characteristics of the nuclear industry demand a highly trained, multigenerational workforce with the appropriate knowledge, attitudes, behaviours, standards and values to sustain the long life cycle of a nuclear facility. In a changing global landscape, meeting these requirements are further challenged by technological innovation, increased mobility both nationally and internationally, changing demographics, and workforce losses due to people changing career aspirations and taking advantage of greater equality of opportunity. This creates a particularly demanding situation for human resource management (HRM) in the nuclear industry.

Formulating and implementing a suitable HRM strategy is critical in order to address these challenges and manage human resources in the nuclear energy field while ensuring safety, security, non-proliferation and performance requirements. A suitable HRM strategy, together with sufficient competent resources and effective processes and procedures, is central to safety and business performance and needs to be included as an integral part of an organization's management system.

This publication aims to provide guidance both at the level of management of individual employees through the plant life cycle and at the organizational level. It elaborates on ten key human resources (HR) processes concerning the management of individual employees as well as the four broader organizational issues — organizational and safety culture, stakeholder engagement, diversity and inclusion, and change management — to which the processes relate.

The publication underscores the importance of the HR function working in tandem with the organization's management functions to ensure success, and as such is intended to support a variety of users including line managers, HR professionals and senior managers.

The IAEA gratefully acknowledges the work of the contributors to the drafting and review of this publication, particularly B. Molloy (Ireland), S. Mortin (United Kingdom) and D. Palmer (United Kingdom). The IAEA officers responsible for this publication were M. Van Sickle of the Division of Nuclear Power and D. Drury of the Division of Planning, Information and Knowledge Management.

CONTENTS

1. INTRODUCTION

1.1. BACKGROUND

The nuclear energy industry involves organizations that either directly or indirectly support the application of nuclear energy for peaceful purposes. The life cycle of typical nuclear facilities can be very long — for individual nuclear facilities this can be over 100 years when considering the phases of planning, siting, designing, constructing, commissioning, operating and decommissioning. The life cycle has an even longer time frame when considering the need to dispose of long lived nuclear waste.

This publication provides guidance regarding the management of human resources for organizations who are involved in the field of nuclear energy. These organizations include the following:

- Operating organizations;
- Regulatory bodies;
- Technical support organizations;
- Research and development (R&D) organizations;
- Fuel and waste transport organizations;
- Educational and training institutions;
- Professional organizations;
- Government agencies;
- Suppliers.

In this publication, these groups are referred to as nuclear organizations.

1.2. OBJECTIVE

The objective of this publication is to provide guidance in the effective management of human resources in the field of nuclear energy to support high levels of safety, security, non-proliferation and personnel and plant performance.

Guidance provided here, describing good practices, represents expert opinion but does not constitute recommendations made on the basis of a consensus of Member States.

1.3. SCOPE

This publication is aimed at a variety of users for different purposes including the following personnel:

- Senior managers, to assist them in leading the establishment of an HRM strategy, and systems and processes needed to ensure a competent, high performing and sustainable workforce.
- Human resources (HR) professionals, to help them develop an appropriate HRM strategy as well as the necessary subsequent processes, which are endorsed by the senior managers. It will also help HR professionals better understand some of the specifics of human resource management in the field of nuclear energy.
- Line managers, to help them understand their role in human resource management and their responsibility for the development, competence and performance of their staff in particular.

1.4. STRUCTURE

In addition to the introduction, this guide contains ten sections containing information for managing human resources across the employee working life cycle process and four sections addressing organizational aspects related to human resource management (HRM). Specifically, Section 2 explains the key elements of HRM and associated processes, the challenges of HRM to nuclear organizations and identifies the key roles of both HRM professionals and company management and the importance of employee relations; Sections 3 to 12 cover each of the main HRM processes; and Sections 13 to 16 address the organizational aspects related to HRM.

2. HUMAN RESOURCE MANAGEMENT IN THE NUCLEAR ENERGY INDUSTRY

As in other sectors, the nuclear workforce is facing many challenges, such as developments in technology, increased mobility both nationally and internationally, changing demographics, changing career aspirations and greater equality of opportunity.

The situation is made even more challenging because, in many Member States, careers in science, technology, engineering and mathematics (often referred to as STEM subjects) generally, and in the nuclear industry specifically, are not seen as being as attractive as other areas such as information technology, media or business.

Additionally, many of those currently working in the nuclear energy industry are retiring, or approaching retirement, and attrition of competent personnel is a significant risk. With the expansion of the use of nuclear power, the number of industry personnel will need to expand significantly. Thus, the industry will have to recruit a large number of suitably educated people and to provide them with the training and experience needed to ensure they are competent and qualified for their roles. Competence includes a combination of knowledge, skills and attitudes (KSAs) in a particular field which, when acquired, allows a person to perform a job or task according to identified standards. Competence (competency) can be developed through a combination of education, experience and training. Personnel are generally only considered competent for a job or role when, in addition to the required knowledge and skills, they have received the specific training and experience required for that role.

Member States and nuclear organizations need to recognize the importance of establishing and implementing an HRM strategy which, together with effective processes, will play a key role in maintaining and improving safety and business performance.

It is important that the correct HRM strategy is in place, together with the right level of competent resources, effective processes and procedures, to support the needs of the organization. These processes and procedures form an integral part of any organization's management system.

2.1. KEY ELEMENTS OF HUMAN RESOURCE MANAGEMENT

In general, the goal of the HRM strategy in any organization is to recruit, retain and motivate a competent workforce, while also retaining the flexibility to vary the size and skillset of that workforce according to the changing needs of the organization. This necessitates the establishment of an integrated HRM strategy, as part of the organization's overall management system.

Effective HRM includes several interacting elements, as shown in Fig. 1. The processes shown on the outer circle relate to the management of the individuals in the organization and address the following aspects of HR:

— Identifying the HRM requirements or workforce planning, including the gaps that need to be addressed;

— Identifying where staff can be recruited from and managing the recruitment process;
— Inducting people into the organization so that they understand the company vision, values, organization, etc.;
— Identifying and providing the required training so that personnel can fulfil their job requirements and develop their potential;
— Managing the remuneration and rewards packages to attract, motivate and retain staff;
— Implementing a performance management process with regular performance reviews;
— Ensuring a succession management process and a process to recognize and develop talented individuals are in place;
— Ensuring a healthy workforce, engaged in the success of the organization;
— Managing employee turnover and retirement.

FIG. 1. Elements of human resource management.

The processes shown on the inner circle relate more to organizational aspects and address the following aspects of HR:

— Organizational and safety culture, particularly important for the nuclear industry, where openness and a priority for safety are essential;
— Stakeholder engagement, the need to engage with all stakeholders in an effective manner;
— Diversity and inclusion, the need to manage and optimize a diverse workforce, this area is of increasing importance as the industry becomes more globally mobile;
— Change and knowledge management, organizational change and the role that people have in ensuring its effectiveness and the importance of knowledge management during and after the organizational change.

2.2. CHALLENGES FOR HUMAN RESOURCE MANAGEMENT IN THE NUCLEAR ENERGY INDUSTRY

In general terms, the HR processes and practices in the nuclear energy industry and nuclear organizations are similar to many of those found in other industries. However, the combination of specific requirements which apply to the industry means that managing HR in the nuclear field is more challenging than many other sectors. Some of those specific requirements include the following:

— Working with nuclear and radiological materials necessitates a safety and security culture that encourages people to have the appropriate attitudes, behaviours, standards and values. This, necessarily, reduces the numbers of candidates that are suitable for consideration in the field.
— Organizations can be staffed with people from different backgrounds and national and organizational cultures. The workforce involved in nuclear organizations, particularly those in the nuclear power industry, is becoming increasingly international. This brings about a different set of challenges in terms of different cultures, language barriers, different working methods, staff mobility and different safety and security standards. Work processes, training methods and retention strategies need to account for these differences.
— The complex technology means that additional job-specific training and experience requirements for some positions can add several years to the overall employee development process. Once trained, there is still a requirement to maintain qualifications using a systematic approach to training (SAT) to maintain competence, which involves considerable

time and resources. This needs to be accounted for in the recruitment and retention policies of the organization.

— Nuclear facilities will often have at least a 100-year lifetime from planning through construction, commissioning and operation to decommissioning. The implications for the workforce are that people from several generations will be responsible for these facilities over their lifetimes. Each generation may have different norms, expectations and motivating factors that need to be considered and addressed in planning and implementing programmes to recruit, develop and retain them. It is also important to address the management of knowledge over a facility's lifetime.
— Nuclear energy requires a high level of oversight. Individuals need to be held responsible for their actions (referred to as nuclear professionalism in some countries) but at the same time there is a need for oversight to assure compliance with rules and requirements.
— In some countries, much of the workforce is retiring or nearing retirement. This means that the pool of expertise available will gradually reduce over time and create even more demand for competent resources. This will also create multi-generational workforces, which could need new approaches to learning and training.

For the sake of completeness, this publication addresses all elements of HRM, recognizing that many aspects are not specific to the nuclear field.

2.3. LINK BETWEEN THE HR FUNCTION AND MANAGEMENT OF THE ORGANIZATION

The role of the HR function is to support managers in acquiring and managing competent staff to enable them to deliver the organization's goals and objectives.

The organization's management and the HR function need to work in tandem. Both have a key role in ensuring success. Some examples of this are shown in Table 1.

As part of the management system, HRM itself will be addressed by the appropriate supporting employment policies, standards, processes, procedures and guides that will help contribute to the organization successfully achieving its vision, mission and objectives within applicable statutory requirements.

TABLE 1. RELATIONSHIP BETWEEN THE MANAGEMENT OF THE ORGANIZATION AND THE HR FUNCTION

Management of the organization	Contribution of the HR function
Mission and business plan	Senior HR staff participates in strategic planning and decision making activities to ensure that the HRM strategy aligns with, and supports, delivery of the organization's objectives. Their participation also enables senior HR staff to highlight any HR challenges related to the strategy (e.g. outdated organizational structures and responsibilities, insufficient resources, training requirements related to organizational changes). The HRM strategy and processes need to align with the business environment and the organization's mission.
Performance	Processes to manage HR are established and key performance indicators are used to measure and increase their effectiveness. HR processes support managers in ensuring the performance and development of staff to achieve the organization's goals and objectives.
Structure	The roles, functions, responsibilities and job profiles of all organizational units and personnel are clearly defined, as are interfaces between units and with external organizations.
People	All employees are appropriately qualified for their jobs, are engaged in achieving the organization's goals and objectives and are motivated to achieve them. Criteria to assure the competence of contractor personnel are clearly defined and implemented.
Communications	Good communication is key to success. HR and management both have an important role. There is regular and timely communication through line managers and other channels. Communications will reinforce the vision, values, culture and objectives of the organization; allow employees to give upward feedback either informally or through employee engagement surveys, etc. (this could include feedback on performance, ideas for improvement and comments on their general working conditions); and provide regular feedback on organizational performance, organizational changes and key management decisions that affect staff.

TABLE 1. RELATIONSHIP BETWEEN THE MANAGEMENT OF THE ORGANIZATION AND THE HR FUNCTION (cont.)

Management of the organization	Contribution of the HR function
HR technology	The best HR systems are usually integrated with the main business system, support data protection and minimize the need for onerous data input. An integrated management system usually saves time, provides a single set of data and increases accuracy and facilitate procedural compliance.

2.4. ROLE OF EMPLOYEE RELATIONS

Employee relations is the term used to describe the relationship between employers and employees. Within nuclear organizations the employee relations strategy needs to take a long term approach to supporting the organizational objectives. It will focus on both individual and collective relationships in the workplace, with an increasing emphasis on helping line managers establish trust based relationships with employees and reducing potential workplace conflict. A positive climate of employee relations, with high levels of employee involvement, commitment and engagement, can improve organizational outcomes as well as contribute to employee well-being.

Employee relations remains an important concept for organizations:

— Labour unions have an important role in many organizations. This is partly through the existence of institutions of collective consultation, reinforced by continued reliance in many cases on industry-level consultations and negotiations (e.g. annual salary increases) and the emphasis on working in ‘partnership’.
— Managers need to develop the right competence, including the correct behavioural as well as technical skills, to be effective people managers, essential to a successful employment relationship.
— Organizations need to train and support line managers in areas such as teamwork and change management as the basis for establishing and maintaining motivation and commitment, which is a critical role for managers. Managing the employment relationship is a key part of the line managers’ role and their competence in this area is crucial and is a strategic issue.

— HR practitioners associated with employee relations are usually involved with drafting and advising on company policies (e.g. grievance or disciplinary), especially those which involve national legislation.

3. WORKFORCE PLANNING

The lifetime of a new nuclear facility, from its design and construction to the completion of its decommissioning, could be 100 years or more (10–15 years planning and construction, 60 plus years' operating life, 25 plus years until decommissioning is completed, depending upon the approach selected). Thus, people from three or more different generations can be responsible for the facility. This characteristic, along with the lack of growth of the industry in many countries, the high standards required of nuclear industry personnel, and the long lead times needed to prepare the personnel for their demanding assignments, mean that anticipating HR needs is particularly important for the nuclear industry.

Also, many of the current nuclear facilities were put in place in the 1970s and 1980s, and the following years saw very little growth in the industry. Thus, the age profile for these facilities is unusual, generally having an older workforce than other major industrial facilities. As many of these facilities are considering life extension up to 50 years or even beyond, this necessitates the replacement, in a relatively short period of time, of much of the experienced workforce. This situation presents particularly difficult challenges, including maintaining the safety and security culture of the organization and transferring knowledge to the next generation.

Workforce planning is the systematic analysis of the size, type and quality of workforce that an organization is going to need as a function of time to achieve its objectives. It identifies what mix of experience and competencies are expected to be needed and helps ensure that the organization has the right number of people with the right competence, in the right place, at the right time. Workforce planning is an integral part of an organization's overall HRM strategy and needs to be consistent with the organization's vision, mission and business plan. For example, workforce planning will identify who needs to be recruited and when and will identify the need for and nature of training and experience required. Other aspects of the HRM strategy will identify how the right staff can be recruited, how the training can be carried out and how staff can be retained.

The terms 'medium term' and 'long term' are often used in connection with workforce planning and these terms are explained as follows:

- Medium term has a time frame of between one and three years and considers the people that will be required to deliver the organization's objectives, taking into account the budget of the next financial year. It will also consider the necessary training and competency development of staff.
- Long term has a time frame of between three and ten years and considers what people are required to deliver the longer term organizational strategy, where and when these people will be required and what skills, qualifications and competencies are needed. It is important that workforce planning is aligned with the longer term organizational planning process.

3.1. CRITICAL STEPS OF WORKFORCE PLANNING

The critical steps of workforce planning are:

- Analysing the future workforce requirements and developing specifications for the competencies, numbers and locations of employees and managers needed to accomplish the organization's mission, goals and objectives. This information will need to be developed in conjunction with the organization's strategic plans and budgetary requirements. Analyses also need to consider current and desired age profiles.
- Assessing the current workforce and determining what the current workforce resources are and how they will evolve over time through turnover (retirements, etc.).
- Identifying and determining what gaps will exist between current and projected workforce needs including the identification of current and emerging skill requirements. These requirements then form the basis of the development of a recruitment, retention and training strategy as discussed in later sections. They will also be a basis of wider company strategies such as restructuring and outsourcing.
- Identifying lead times for recruiting, hiring and training and transferring knowledge. The timing of these requirements needs to be included in the workforce plan to ensure timely development of personnel to meet the 'just in time' delivery of skilled personnel for the future or for the effective transfer of knowledge.

The workforce plan is a 'living' document that requires scheduled reviews and updates. This becomes more important in a time of organizational change, or when the labour markets change, and the actual workforce attrition differs from what was planned. In this regard, it is advisable to use an appropriate IT tool in order for the plan to be easily updated and to stay 'current'. The plan will include the identification of risks and risk mitigation measures for unexpected conditions.

The benefits of workforce planning include the following elements:

- Providing a systematic structure within which to evaluate alternative organizational designs from an HR perspective;
- Identifying expected gaps between the competence of the existing workforce and those that will be needed based on plans for the future;
- Assisting in developing a long term strategy, including outsourcing, for the recruitment, training and employment of future employees;
- Addressing the replacement of employees through attrition in critical knowledge areas and competence gaps, which can occur as a result of process improvements, technology advancements and changing organizational requirements;
- Assessing the extent to which the available workforce can be effectively utilized to support the life cycle of a facility, from commissioning through major upgrades to decommissioning;
- Identifying potential gaps in the national education and training programmes and infrastructure, which could need government intervention and support.

3.2. OUTSOURCING STRATEGY

The use of personnel from third party organizations is a key feature of most workforce plans in the nuclear energy field. The number and competence of the permanent staff required depends on the outsourcing strategy of the organization. For example, on some nuclear power plants maintenance and some technical support functions have been performed through a contract with the original supplier of the plant systems or components or with another contractor. In the case of the regulator, some assessment and inspection services can be outsourced.

The range of outsourced services will be determined by the following:

- The types of services are required — whether they are a one-off, temporary or long term basis;
- The difficulty of maintaining competence for infrequent specialist tasks;
- The availability of internal and external skills and resources;
- The availability and cost of external services;
- Legal and regulatory requirements;
- The ability of the organization to hire and retain personnel internally;

— The minimum staffing that will be required to cope with accident conditions and scenarios;
— Other considerations, including the risk management approach of the organization.

Whatever the situation, a nuclear facility licensee can never delegate its responsibility for safety and must maintain sufficient competent staff to effectively manage any outsourced activity.

4. CANDIDATE SOURCING

The workforce plan will detail the required number of people, their knowledge, skills and experience and when they are needed. Having defined this, the next step is to answer the following questions:

— Where can suitable candidates be found? (Candidate sources and pipelines.)
— How can those candidates be reached? (Channels to reach candidates.)
— What will attract these people to join the organization? (Recruitment of candidates.)
— How can an organization enhance its reputation as an employer? (Employer branding.)[1]

4.1. CANDIDATE SOURCES AND PIPELINES

Candidate sources and pipelines can be broken down into two categories: (1) those with no experience, coming straight from school, technical/vocational training institutions or universities and (2) those coming from related industries with some relevant skills and experience (e.g. electricity generation, nuclear research facilities, nuclear navy, military or high risk/high safety industries) or with specific transferrable skills (e.g. legal, financial, procurement, HRM).

The first step in working with both of these categories is to map the competence requirements of the organization to be able to compare these with the potential candidate sources.

---

[1] Employer branding is the process of managing and influencing your reputation as an employer among job seekers, employees and key stakeholders. It encompasses everything you do to position your organization as an employer of choice.

4.1.1. Candidate sources and pipelines from education and training institutions

Through working closely with education and training organizations, it is possible to influence the curricula to develop people with KSAs which will fit better with the needs of the workforce plan. Examples include partnerships with schools or universities to develop courses that equip people with the knowledge and skills needed. These organizations include schools (~ ages 5 to 18); technical and vocational schools/colleges (~ages 16 to 20); and universities (~ages 18 to 22 or older).

To facilitate and improve candidate pipelines and sources, nuclear industry managers and key stakeholders (including government) will establish relationships and partnerships with relevant academic and training institutions. From an industry perspective, the role of education is to provide people with the capabilities to become competent professionals and to prepare them for industry and job specific training programmes. However, education and training institutions are typically preparing students for a wide range of careers and could be unaware of the needs of specific sectors such as the nuclear sector. The nuclear industry needs to therefore work with the education and training sector to develop programmes which are aligned with the nuclear industry's professional and technical standards and requirements. Mechanisms to ensure the quality of education and training programmes need to be established and could include accreditation, standardization, internships, apprenticeships and other cooperative programmes for students. They could also involve nuclear industry managers and leading specialists teaching at education and training institutions. In some Member States, university engineering programmes are accredited by national professional engineering organizations, making these bodies important partners for the nuclear industry.

Member States can also have government sponsored or funded vocational education and training programmes operated by academic institutions that produce qualified technicians. These technicians typically have a basic set of qualifications and practical experience in their technical field. In this case, a nuclear industry organization needs only to provide facility specific and nuclear fundamentals training after the individual joins the organization.

It is important that government agencies, industry and academia collaborate nationally and internationally to create a framework to support education and training for the nuclear energy sector. This includes considering funding and planning for nuclear R&D being integrated with funding for education. Similarly, organizations funding nuclear R&D need to ensure that education and training aspects are included as a component of the research activities. Networking of academic institutions is a key strategy for capacity building and making better use of available educational resources.

4.1.2. Candidate sources and pipelines from industry

Following the analysis of the KSAs required by the organization, it is possible to identify those industries and organizations whose workforce can already have some of the competences required for the nuclear industry. These sources include the following organizations:

— Organizations in the nuclear industry (e.g. research reactors, nuclear medicine);
— Organizations in other adjacent or relevant industries (e.g. energy, rail, petrochemicals, aviation or mega infrastructure);
— Military, especially navy personnel with nuclear propulsion experience.

This may mean competing against other potential employers to recruit these personnel and this will need to be reflected in the HR recruitment and retention strategy. However, even in these areas it may be possible to develop strategic partnerships with some of these organizations. For example, in the case of nuclear navy personnel, they typically retire at a relatively young age from active service and their employers are often trying to help them to find second careers. Care needs to be taken when recruiting candidates with previous experience from other industries, as they could bring with them attitudes and behaviours that are not compatible with the safety culture in the nuclear industry.

4.2. CHANNELS TO REACH CANDIDATES

When the sources and pipelines of relevant candidates are identified, the next consideration is how to reach and communicate with these people. A number of mechanisms are available, and these will be considered in terms of their suitability and cost effectiveness.

4.2.1. Internal recruitment function

Establishing an internal recruitment team to identify and screen internal and external candidates for job roles. Typically, they place advertisements, use social media (e.g. LinkedIn) and build a database of potential candidates for the organization to interview. This can be an efficient and cost effective method, where the volume of recruitment is sufficient to justify the fixed costs.

4.2.2. Recruitment agencies

The alternative to an internal recruitment function is to outsource to an external agency. It could be that all recruitment is outsourced, or that some is managed internally and the recruitment for more specialized roles is outsourced. The cost of using recruitment agencies varies but is typically a percentage of the first year salary. The quality of recruitment agencies varies, and it is advisable to partner with those agencies that have a clear track record in recruiting the type of people the organization needs and have credentials in the nuclear energy industry.

4.2.3. Executive search companies

For more senior roles an executive search service can be the most appropriate option. These searches are designed to thoroughly assess the entire field of potential candidates before returning a short list and are suited to senior managerial or technical roles. The cost for this service is again typically based on a percentage of first year salary. It is common for part payment to be made in advance and part on completion of the hire.

4.2.4. Partnerships and alliances

If strategic alliances are formed with organizations such as universities or schools, as described in subsection 4.1, these relationships can facilitate the flow of suitable candidates to the organization.

4.2.5. Referrals from existing employees

Some organizations operate schemes whereby their employees are encouraged to make referrals of suitable candidates to recruit from their own networks.

4.3. ATTRACTING CANDIDATES

Having identified the sources of candidates, and channels to reach them, it is essential to understand what will attract the right candidates. The following questions are key:

— What are the motivations and needs of the target candidates? This information can be gained by speaking with a sample of target candidates directly or using partners such as recruitment agencies to gain an understanding of what the target candidates are looking for.

- How can the organization match the motivations of target candidates? This means considering salary expectations, benefits, working conditions, flexibility, working location, learning and development opportunities, career progression, etc.
- How does the organization compare against competing employers? It is important to benchmark what the organization can offer against competing employers, which could be in nuclear or other industries. This also allows the identification of differentiators which can be emphasized to increase candidate attraction.

Section 8 discusses remuneration packages in more detail.

4.4. EMPLOYER BRANDING

Employer branding is important in enhancing the reputation of nuclear organizations from other competing industries when it comes to attracting candidates. The employer brand is directly connected to the needs, motivations and preferences of target candidates; competitive advantage can be achieved by understanding what the organization can offer that is more attractive for potential applicants compared to what competing organizations can offer.

Various methods can be used to ensure that employer branding is effectively communicating the message to target candidates, such as social media, web sites, articles and employment fairs.

Finally, it is essential that the 'employer value proposition' (defined as a set of associations and offerings provided by an organization in return for the skills, capabilities and experiences an employee brings to the organization) [1] and resulting employer brand closely align with the organizational culture and the employee experience. The ultimate goal of a strong employer brand is to recruit and retain high quality candidates for the organization. Once these candidates become employees, they will expect to live the brand they experienced during the recruitment process. New hires attracted to the organization hopefully will quickly become engaged and passionate employees, and then become part of the employer brand themselves.

5. RECRUITMENT

Having identified potential candidate sources, the recruitment process will be owned and implemented jointly by HR professionals and hiring managers.

An effective recruitment process will include the following elements:

— Role definition;
— Generating applicants;
— Application, interview and selection;
— Offer and acceptance;
— Onboarding and mobilization;
— Documentation, evaluation and systems.

Each of these are addressed in the following sections.

Consideration may be given to taking a graded approach for recruitment; positions affecting nuclear safety requiring more attention than other positions.

5.1. ROLE DEFINITION

5.1.1. Job analysis

Before recruiting for a new or existing position, it is important to consider not only the duties involved in the job and the skills and experience required, but also the job's purpose, the outputs required by the job holder, how it fits into the organization's structure and any potential changes in the job role. An example template of a job analysis form can be found in Annex I.

5.1.2. Job description

The job analysis provides the information needed to create the job description, which can include the following details:

— Contextual information about the organization, department or project in which the role resides, including reporting lines and any direct reports;
— The purpose, objectives and main responsibilities of the role;
— A clear overview of the technical and behavioural competence requirements;
— Any specific requirements such as education, accreditations, job-related medical requirements and languages;

— Practical information around the location, working hours and flexibility of the role;
— Pay and benefits.

5.2. GENERATING APPLICANTS

5.2.1. Internal applicants

It is important to explore the internal talent pool when recruiting for a position. Recruiting internally provides opportunities for promotion, development and career progression, increases employee engagement and retention and supports succession planning. Internal promotion results in the vacancy appearing at a lower level in the organization, making it easier and more cost effective to replace externally. The long training periods needed for nuclear facility personnel can also cause organizations to favour promotion from within over recruiting externally. It is common to launch any recruitment process by advertising the role internally for a period of time (e.g. two to four weeks) before considering external applicants.

5.2.2. External applicants

There are many options for generating interest from individuals outside the organization as detailed in the previous section of this document. The most common ways for attracting candidates include the employer's web site, internet job boards, recruitment agencies and professional networking sites. Applicants expect to be able to search and apply for jobs on-line and this means that organizations need to pay attention to their corporate web site and their employer brand.

It is important that advertisements be clear and written in positive language to encourage target candidates to apply. Care needs to be taken to ensure that the wording and description of the job is accurate, compliant with legislation and does not inadvertently deter candidates from applying.

5.3. APPLICATION, INTERVIEW AND SELECTION PROCESS

5.3.1. Managing applications

A process needs to be established to receive and evaluate candidate applications. The process will designate the responsibilities for evaluating the applications and providing feedback to applicants in a timely manner. Regular communication with the applicants is extremely important throughout this process.

Applications normally comprise a resume or curriculum vitae (CV), along with a cover letter or a completed application form. The use of application forms allows candidates to present information in a consistent format, which will facilitate the collection of information from job applicants in a systematic way. This in turn will allow the recruiter to objectively assess the candidate's suitability for the job. To comply with anti-discrimination laws, application forms may need to be offered in different formats. Additionally, all applications need to be treated confidentially and restricted to those individuals involved in the recruitment process. Prompt acknowledgement of the receipt of applications is a good practice and projects a positive image of the organization.

If there are any requirements that applicants must meet for the job position, such as security clearance(s) or a working visa, it is important that these requirements are clearly communicated and that the application process is designed to ensure that only qualified applicants are selected for interview.

There needs to be clear criteria for the inclusion of applicants in a shortlist based on the requirements identified in the job description, both to ensure that appropriate candidates are selected for interview and to protect the organization from any claims of bias or discrimination.

5.3.2. Interview and selection

The objective of the interview and selection process is to identify the most suitable candidate for the requirements of the job position. During this process, it is important to continue regular and positive communication with applicants and to provide them with all of the relevant information to support their application. It could be an HR professional or a hiring manager that owns the process and takes responsibility for the candidate experience.

Selection methods are designed to assess the extent to which candidates have values consistent with those of the organization, such as honesty and respect for others. Differences in national laws and culture regarding employment and individual privacy dictate which selection tests and methods are suitable for a particular organization. In some Member States, selection methods are required by law to be based only on job requirements and include the following approaches:

— Telephone interviews;
— Face-to-face interviews in person or using video links;
— Panel interviews;
— Assessment centres for individuals or groups;
— Tests, such as psychometric, technical, etc.;

— Work samples, such as inviting applicants to the working site for a short period.

Taking time to consider the most appropriate selection methods and then ensuring that those carrying out the selection are suitably trained are important and have direct influences on the quality of hiring decisions and retention. Considering the cost associated with training for specialist nuclear roles, it is highly cost effective to invest in the recruitment and selection phase in order to select candidates with the best chance of successfully completing their training programmes. Identification of attitudes and behaviours consistent with a strong nuclear safety culture will be part of the selection process to ensure the success of candidates in the organization. Panel interviews are a common selection method and are beneficial because it provides varying perspectives, helps to eliminate biases, and benefits the department by involving interdepartmental managers and stakeholders in the hiring decision.

5.4. OFFER, ACCEPTANCE AND FITNESS FOR DUTY

Prior to making an offer of employment, organizations are responsible for verifying that applicants have the right to work in the country where the worksite is and have the appropriate qualifications or credentials. The recruitment policy needs to state clearly how references will be used, when in the recruitment process they will be taken and what kind of references will be necessary (e.g. from current or former employers). These rules need to be applied consistently, and candidates need to always be informed of the procedure for obtaining references. References are most often sought following a 'provisional offer' given to an applicant but can also be used to help decide between similarly suitable candidates.

Identification, recruitment and hiring of staff to perform roles that require an additional level of reliability and trust, often include a thorough review of character and trustworthiness. This can include criminal background checks, review of financial records, prior work history verification, as well as drug and alcohol and psychometric testing to ensure that the preferred candidate meets the organization's requirements. This reinforces the organization's commitment to safety and security by ensuring that staff are highly reliable, trustworthy and are performing their duties to the best of their ability. This process is often referred to as fitness-for-duty, a human reliability program or personnel reliability program [2].

There needs to be a well-defined process for issuing a formal job offer and managing the documentation and communications during the offer and acceptance stage. The process will also include the final determination of pay and benefits and have clear communication channels for applicants to ask questions or seek clarification. All documents need to be saved and recorded, with signatures gathered when required.

5.5. ONBOARDING AND MOBILIZATION

The onboarding and mobilization process describes what happens from the point that an offer is accepted until the candidate joins the organization. This process includes the following actions:

- — Providing information about the organization and the local area;
- — Supporting the candidate and their family if they are relocating, in terms of financial support as well as advice and practical support;
- — Communicating clearly through a process that leaves the candidate feeling supported and gives them a point of contact for any questions;
- — Managing effectively requirements including medical tests, security clearance and immigration processes.

5.6. DOCUMENTATION, EVALUATION AND SYSTEMS

The documentation necessary to support the recruitment process needs to be well defined, accurate, comprehensive and securely archived, with access limited to authorized staff. Information needs to be kept for sufficient time so that archives can be used to resolve complaints; the length of time required to keep records will vary in each Member State. All of the information gathered needs to be stored in line with data protection legislation.

It is good practice to monitor applications and recruitment decisions to ensure equality of opportunity (e.g. gender equality). Where issues are highlighted in the process, corrective action needs to be taken.

Using metrics such as cost per hire, candidate experience ratings and time to hire (e.g. from registering the vacancy until a job offer is made) can also provide insight into the effectiveness of the recruitment process. Having a suitable and effective IT system will help support and enhance the recruitment process and facilitate timely and relevant key performance indicators.

There are many software systems designed specifically to support the recruitment process, where applicant CVs can be stored, the recruitment process can be managed, and all communications can be sent and stored. Most organizations have implemented such systems or designed their own.

6. INDUCTION AND ORIENTATION

Induction refers to the process where new employees are introduced to the organization, their jobs and the working environment.

Every nuclear organization needs to have a well-considered induction programme that provides new employees with a positive experience of the organization and highlights the importance of the safety and security culture from the first day. It will provide all the information that new employees need, without overwhelming or diverting them from the essential process of integrating into their team.

Research demonstrates that induction programmes benefit both employers and employees. For organizations these include improving the person–job fit, reducing turnover and absenteeism and increasing employee commitment and job satisfaction.

6.1. PURPOSE

For employees, starting a new role in a new organization can be an anxious time and an induction programme enables them to understand more about the organization, the safety and security culture, their roles, ways of working and to meet new colleagues.

For nuclear organizations the purpose of induction is even more important to ensure that new employees, including contractors, understand the terminology used in the nuclear industry. These new employees need to understand the standards and behaviours expected of them, especially expectations regarding nuclear safety and in emergency situations.

'Nuclear professionalism' is a behavioural standard and expectation and it is not only about defining the behaviours required but also putting in place a set of supporting tools that are embedded into an organization's processes and management practices. These behaviours need to be embedded from the first day in order to achieve the expected standards of professionalism.

In recent years, many employers have implemented 'shadow-training' for new employees to foster a comprehensive understanding of the facility before they are assigned to a particular post.

Increasingly, nuclear organizations are working in a more networked and globalized way, and often much of the induction process is the same for employees, contractors and consultants and can be used to cover access to multiple sites within one organization.

6.2. INDUCTION PROCESS

The length and nature of the induction process depends on the type and level of job role, the background of the new employee, and the size and nature of the organization. Regardless of the organization size, the key areas that induction can include are detailed in Table 2.

TABLE 2. KEY AREAS TO INCLUDE IN AN INDUCTION PROCESS

Key area	Details
1. Physical orientation	An escorted tour of the department/facility and an introduction to fellow employees.
2. Health, safety and security	The safety culture characteristics and behaviour expectations, the security culture characteristics and behaviour expectations, day-to-day guidance in local procedures, emergency arrangements and health and safety information.
3. Overview of the organization	Presentation on the specifics of nuclear energy, the safety and security issues that arise from it and the importance of the role of the employee and their behaviour. Organization history, strategy and objectives, products and services, and culture. Details of benefits and memberships. The overview is sometimes offered via a video or digital learning.
4. Team and job	Explanations of the departmental organization and the requirements of the job.
5. Managing performance	Expectations of the new employee, the purpose and operation of any probationary period and the performance management system.
6. Training and development	Information on available learning, training and development services, instructions regarding developing a personalized development plan, along with details of other sources of information during induction such as the company intranet or interactive learning facilities.
7. Meeting key people	Introductory one-to-one meeting with key members of the organization. This ensures new recruits can meet more people in the organization, understand their roles and how they can work together. In organizations where the workforce is dispersed across multiple locations, the use of digital tools can allow employees to meet colleagues in other areas of the business.

For operational nuclear facilities, a central element of induction, often referred to as site access training or general employee training, is usually given to all staff and contractors. This training can include such topics as site access procedures (including for restricted areas), emergency procedures (including alarms) and overviews of the organization, functions and responsibilities. Again, the length of this training varies on the person and their role but is typically between one half day and three days.

Nuclear organizations need to also pay attention to employee experience before the first day of employment; for example, ensuring pre-employment communications are engaging, as well as using social network sites to put new recruits in touch with each other. This is particularly useful for recruits who recently graduated or are apprentices.

It is also important that the process continues into employment. A 'buddy' system or mentoring programme incorporating experienced staff members to provide more informal support and help new employees settle into the organization could be used for an initial period.

6.3. BENEFITS OF AN EFFECTIVE INDUCTION PROCESS

A well-designed induction process results in positive first experiences. This means that the employee integrates into their team, becomes productive quickly and works to their highest potential.

An effective induction process includes the following benefits:

- New employees understand the organization, their role and the culture more quickly;
- Reduction in recruitment costs due to low turnover rates in the first twelve months of employment;
- Positive experience for line managers;
- A sense of belonging to the organization at the earliest opportunity;
- Improved reputation and employer brand.

The induction process needs to be monitored to determine whether it is meeting the needs of the new employees and the organization. Monitoring can include opportunities for feedback at the end of the induction process, performance indicators and exit interviews particularly from those who leave within the first twelve months of employment.

6.4. DELIVERY METHODS FOR EFFECTIVE INDUCTION

For a large organization, the induction process is likely to be a combination of one-to-one discussions, more formal group presentations using a modular approach and/or using an induction course (some or all of which could be e-learning). There are advantages and disadvantages of using an induction course as listed in Table 3.

There are a number of key principles that are important to understand when managing any effective induction process:

— HR and line managers need to avoid providing too much, too soon — a new employee must not be overwhelmed by a mass of information on the first day. They need to keep it simple and relevant.
— Presentations need to be given at an appropriate level — where possible, presentations need to be tailored to consider prior knowledge of new employees.
— HR and line managers need to provide all the information through a shared process. However, subject matter experts can be used where appropriate.

TABLE 3. ADVANTAGES AND DISADVANTAGES OF USING AN INDUCTION COURSE

Advantages	Disadvantages
Saves inductors and managers time by dealing with a group rather than several individuals; the use of digital tools to share information can be useful where new recruits are globally dispersed.	Contains a range of subjects that are unlikely to appeal to a cross-functional and mixed ability group of new employees.
Ensures that all new recruits are given a consistent positive message portraying a clear employer brand, values and culture.	Could take place several weeks, or even months, after the inductee joins the organization, which disrupts integration into the work team and risks information being shared too late in the induction process.
Can employ a range of engaging communication techniques such as group discussions or projects.	Can be less personal and involve managers and HR personnel rather than colleagues and local supervisors.
Enables new recruits to socialize with each other and build cross-functional relationships.	

- — The induction programme needs to generate reasonable expectations and not oversell the job.
- — An induction programme needs to focus not only on administration and compliance but also reflect organizational values and behavioural expectations.
- — Induction content needs to be made available for future reference.
- — An induction programme needs to engage and assure the new employee that they have made the right decision to join the organization.

7. TRAINING AND DEVELOPMENT

Competencies are KSAs in a particular field, which, when acquired, allow a person to perform a job or task to identified standards, including performance and behavioural standards. Competence, or applying competencies, is built and maintained through a combination of education, initial and continuing training and experience, and performance improvement initiatives. Expectations and standards regarding individual competencies and behaviours in a nuclear organization will be defined as part of the management system and will be the standard for all nuclear organizations.

All three domains of competence (KSAs) are important to ensure a high level of performance at work. To acquire various types of knowledge, explicit or tacit, appropriate knowledge management, and education and training needs to be employed. Appropriate behaviours of personnel in nuclear organizations have to be ensured. Due attention needs to be paid to the fact that the required behaviours cannot be achieved solely through education and training. Behaviours also depend on individual characteristics and organizational culture. The behaviour of managers and their ability to be everyday role models for their personnel are crucial factors.

The role of experience in building competence is important and needs to be clearly identified. The acquisition of the right competence could take several years in some areas.

The most effective way to develop a new skill or attitudes is to apply knowledge in the workplace and practice skills in real life situations or, where safety related constraints exist, in a simulator or replica environment. Training activities using simulators, training in workshops and laboratories, or structured in-plant on-the-job training, are effective methods of gaining job experience.

Work on any site needs to operate under a 'zero harm' health and safety policy where everyone on the site is expected to be proactive in dealing with hazards and to help create a harm-free workplace by adopting the appropriate behaviours. In support of this policy, training needs to be provided when undertaking first-time, sensitive, or complex tasks, where simulations are used to practice and validate the tasks before they are carried out in an operational environment. For such tasks, readiness reviews are used to verify the state of preparedness as well as to validate worker training and qualifications. To develop appropriate behaviours in abnormal and accident conditions, and for emergency preparedness, full scope replica simulator training or emergency drills can provide opportunities to gain 'hands-on' practice in a close-to-real work environment.

7.1. SYSTEMATIC APPROACH TO TRAINING (SAT)

The training and qualification of people in nuclear organizations is viewed as a process within the organization's overall management system and needs to be fully integrated into this system.

The SAT needs to be used for attaining and maintaining the competencies of nuclear personnel [3]. It has become the global standard for training in the nuclear industry over many years. SAT consists of five interrelated phases described below and outlined in Fig. 2:

- Analysis: This phase comprises the identification of training needs, tasks or competencies selected for training, and the KSAs required to perform a particular job.
- Design: During this phase, training objectives derived from the training requirements identified during the analysis phase are developed; test items to assess achievement of training objectives are developed; and training objectives are organized into a training programme.
- Development: During this phase, training materials are prepared so that the training objectives can be achieved.
- Implementation: During this phase, training is delivered by using the training materials developed.
- Evaluation: During this phase, all aspects of training effectiveness of both the training programme and training system are evaluated, followed by feedback leading to training and facility performance improvements [3].

The analysis and design phases ensure that training is focused on the necessary job specific competencies. Evaluating the training programme's performance and continual improvement helps maintain the training programmes and keep them up to date and significantly contributes to the quality of training.

The benefits of SAT based training and qualification processes are numerous for both the organization and its personnel. The most important of these benefits is ensuring both the quality and relevance of training. However, for a nuclear

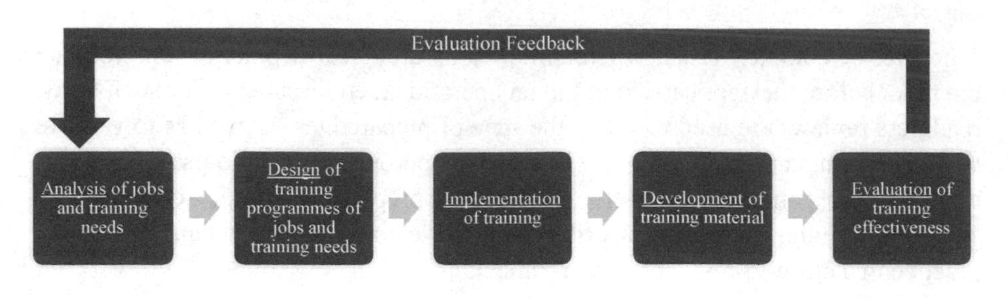

FIG. 2. The five interrelated phases of SAT.

organization to achieve these benefits from SAT, line managers need to believe that training programmes for their personnel 'belong to' or are 'owned by' them.

The implementation of SAT based training requires significant resources. Accordingly, a graded implementation of SAT can be an appropriate option. A graded approach includes the use of different analysis phase techniques or selecting particular training methods, considering the following factors:

— Mission of the organization;
— Life cycle stage of a facility;
— Particular characteristics of the facility or organization;
— Type of position (e.g. maintenance employee, operator, manager);
— Importance of the job relative to safe and reliable performance;
— Magnitude of any hazards involved.

SAT based training has been established as a guideline for nuclear organization training programmes through a variety of mechanisms. In some Member States, SAT based training is a regulatory requirement, while in others it is established through industry initiatives or by an operating organization's policies.

7.2. QUALIFYING AND AUTHORIZING EMPLOYEES

The qualification and authorization of personnel are important elements when ensuring personnel competence. The qualification of personnel needs a rigorous approach and has to be recognized through formal assessment of competence, including demonstration and assessment of its identified elements.

A critical component of qualification and authorization processes is the assessment of whether personnel have achieved the standards needed for satisfactory job performance. Through the SAT analysis and design phases, these standards are identified and included as training objectives. Such standards include not only those related to nuclear technology, but also the behavioural skills needed for satisfactory job performance.

The nuclear industry invests significant resources into conducting assessments of competencies. The following types of tests are used for assessment:

- Written examinations;
- Oral examinations;
- Performance assessments;
- Computer based tests.

Important components in the assessment of competencies include the following:

- Development of tests are based on an analysis of the requirements for competencies (e.g. using job and task or job competency analysis data, training objectives).
- Reliability and validity of tests need to be ensured.
- Results of assessment are analysed to use them in the qualification and authorization processes, and also in improving training (including the process of assessing personnel competencies).

Authorization is the granting of written permission for an individual to perform specified activities in a nuclear facility based on competence, while *qualification* is a formal determination that an individual possesses the education, training and experience specified for a particular job or function. Put simply, all personnel must be qualified for the tasks they undertake; only certain roles may need to be authorized, usually based on national norms or regulatory requirements. Both qualification and authorization are used to permit an individual to work without direct supervision, depending on local and national policies. This authorization can be granted by the organization itself or, for some significant safety roles, by the regulatory body or some independent organization assigned by the regulatory body. In most Member States, only a few positions, most commonly control room operator positions, require formal authorization. However, the methods and practices for authorization vary between Member States. In some Member States, the regulatory body grants the authorization for positions such as nuclear facility operators; in other Member States, the operating organization has the responsibility for granting authorization for various nuclear facility personnel including some management positions. Qualification and authorization requirements need to be clearly identified in job descriptions. For more information on the relationships between SAT based training and authorization see Ref. [4].

7.3. INITIAL AND CONTINUING TRAINING PROGRAMMES

Nuclear organizations normally identify those positions or job families that need the completion of initial and continuing training programmes, developed through the application of SAT, prior to working on tasks without direct supervision. Examples include the following positions or job families:

— Nuclear facility operators;
— Maintenance personnel;
— Health physicists;
— Security personnel;
— Leadership and management personnel;
— Instrumentation and control technicians;
— Nuclear engineers;
— Radiation protection technicians;
— Quality control inspectors;
— Fuel management engineers;
— Waste management technicians;
— Chemistry technicians;
— Electricians;
— Instructors;
— Emergency planning and preparedness personnel;
— Commissioning and test engineers;
— Nuclear research facility experimenters.

Continuing training programmes are provided to ensure that nuclear personnel maintain the competence to perform their assigned tasks, including those they may perform infrequently, such as responding to emergency or abnormal conditions.

One of the characteristics of SAT based training programmes is that, during the analysis and design phases, specific tasks and competencies are identified that need to be included in the continuing training programme for reasons such as the following tasks:

— The task is performed infrequently on the job, and thus proficiency may not be maintained;
— The task is complex and difficult to perform;
— The task is critical to the organization's mission and/or is difficult to perform, thus periodic practice needs to be provided.

This information is also useful in identifying knowledge related to these tasks in areas such as fundamentals and regulatory requirements that will be included in continuing training programmes. Continuing training will also address knowledge regarding, for example, facility modifications and changes in procedures and other documentation.

Event analyses, internal and external operating experience, corrective action programmes and self-assessments all provide information which can identify performance improvement opportunities as well as the content of both initial and, especially, continuing training. Continuing training is an important tool for performance improvement.

Some nuclear facilities have adopted just-in-time (need based) training, particularly for infrequently performed and/or difficult tasks as either an alternative to, or to supplement, continuing training.

Continuing training programmes are also important for professional staff as a tool to keep them up to date regarding new techniques in their professions. Such programmes can include not only the opportunity to participate in formal training courses and workshops, but also assignment to upgrade or modernization projects or nomination to industry/professional events and programmes.

To make initial and continuing training more efficient, individualization of training and appropriate waivers from training are often used. Where waivers are used, these need to be based on competence/performance assessment and not just be given based on time served.

7.4. TRAINING ROLES AND FACILITIES

Senior managers and line managers have an important role to play in sustaining effective training programmes. There are some specific factors that will help to enhance line management ownership and control of training and can be found in Annex II.

Technical competence, positive and demonstrated attitudes towards enhancing the organization's safety culture, instructional capabilities and behavioural skills are all important for instructors. A systematic and reliable process for selection, initial training, performance assessment, qualification and continuing training of instructors needs to be established. Instructors need to possess appropriate competence in pedagogy (teaching), performance evaluation and improvement.

Feedback to the instructors and any external training organization involved in the process, on the basis of evaluation of actual performance of trainees, needs to be provided in a timely and objective manner.

Rotation of personnel between the nuclear facility and the training department is an effective means of maintaining the competence of the training staff; such rotation also strengthens the relationships between the nuclear facility and the training department and cultivates an understanding in the facility personnel of the significance and complexity of training.

Training facilities may belong to the nuclear facilities, or be established at the corporate or industry level, or may serve as (government) training centres on a regional or interregional basis. Sharing training infrastructure can be a cost effective approach. Training centres (either national or regional) will be used more efficiently if considered as centres of excellence for the accumulation of good practices, the training of personnel and sharing experiences.

Although face-to-face classroom instruction is used by almost all organizations, technology based training is gaining in popularity. On-line courses allow organizations to customize learning for individual needs and preferences and provide the ability to measure performance. Virtual reality offers simulated training that mimics employee job duties. Trainees need to be given opportunities, through role playing, case studies, and scenarios that simulate actual conditions, to gain experience before they confront those situations in the workplace.

7.5. LEADERSHIP TRAINING AND DEVELOPMENT

The selection, training, qualification, performance assessment and competence development of leaders and managers is critical to the successful operation of any organization in the nuclear industry.

Some nuclear organizations have established corporate academies that focus on the special needs of the organization in areas such as management, leadership, soft skills and specialized technical knowledge. These corporate academies often have formal or informal relationships with national educational institutions.

'Off the shelf' traditional leadership programmes provide development opportunities but tend to be less effective because they will not be customized to the organization. Ideally, leadership programmes need to be developed based on experience from the wider nuclear industry, combined with cutting edge leadership development practices. These programmes will also benefit from the experience of developing and using accredited technical training programmes, and input from the organization's staff.

The following principles are important to consider when designing any nuclear leadership programmes:

— Nuclear focus: The programmes need to put leadership in a nuclear context and nuclear in a leadership context. While well-established, traditional leadership skills and knowledge can be covered, their application to nuclear safety, nuclear security and organizational improvement needs to be emphasized. The programmes need to be designed using SAT principles, analysing evidence of substandard performance, investigating peer review findings and areas for improvement and interviewing nuclear leaders. By keeping nuclear safety and nuclear security as both the focuses and the main themes throughout, the programmes should satisfy the specific needs of the nuclear industry. This can also help to foster the appropriate culture, or culture change if required, by developing highly effective leaders who all share a common language and the skills to communicate this throughout the workplace.

- Selection process: Programme delegates need to be selected from various departments and levels of the business creating a dynamic and vibrant exchange of experience and the strengthening of cross-functional relationships, with much of the learning coming from the delegates themselves.
- Leadership involvement and support: Where practicable, executive team members and senior nuclear leaders will provide the bulk of the teaching faculty and be trained to deliver many of the programme modules. Alongside professional facilitators with nuclear backgrounds, they bring their career experiences and knowledge into the classroom to share with the next generation of nuclear leaders. Provision of a programme mentor, a senior manager attached to the group throughout the learning process, is essential to ensuring the success of any programme. Line managers also need to be involved — from the nomination of delegates to attend programmes and facilitating comprehensive pre-course briefings and assessments to supporting the action learning work undertaken back in the workplace.
- Action learning and application: Participants will form action learning groups and take their learning back into the workplace. A powerful consequence of this is groups supporting each other and creating lifelong professional relationships as they apply their newfound knowledge and work together on specific challenges.

8. REMUNERATION, BENEFITS AND RETENTION

Remuneration and benefits normally cover all financial and non-financial provisions made to employees, including salary and the wider benefits package (e.g. pensions, paid leave, health coverage).

Remuneration can be divided into two categories:

— Base (or fixed) pay — guaranteed salary paid to employees for doing their work for a contracted period of time;
— Additional variable earnings such as bonus payments or overtime earnings.

Remuneration and benefits are important factors in the process of recruiting and retaining employees in nuclear organizations. There is a range of options available to reward employees and recognize their contributions, each with their own opportunities and risks, but the most effective remuneration and benefit packages will be aligned with the organizational and employees' needs and reflect the organization's objectives and performance. National norms and other industry packages are important considerations as these will determine the relative attractiveness of any package being offered.

When creating a rewards package, it is important that the various elements are integrated so that they support, rather than contradict, one another. Nearly all activities related to the safe and reliable operation of nuclear facilities involve employees working as part of a team, such as part of an operational shift, a maintenance group or a work planning organization. Thus, incentives need to focus not only on individual employee behaviour and competencies, but also on how well employees work as members of a team and how they contribute to the organization's overall performance. Many nuclear industry organizations give all employees a stake in the performance of the organization, offering team based bonuses, and include other related measures to ensure that individual reward does not damage teamwork and cooperation.

8.1. PAY STRUCTURES

Pay structures provide a framework for valuing jobs and understanding how they relate to one another within the organization and to the external labour market. Pay structures may also need to allow for certain additional elements other than basic pay rates, for example, the inclusion of location allowances. There are various approaches to setting pay levels or ranges. Job evaluation (a systematic way of determining the value/worth of a job in relation to other jobs in an organization) is one example for setting pay rates with market pricing. It can also be used to determine the external value of a job. Where market rates are used, employers need to determine where to pitch their 'in-house' rates (for example, at the median or upper quartile).

As mobility is an important factor in encouraging movement within the nuclear industry, location allowances may need to reflect the local cost of living if there are significant differences between locations. Typically, living costs increase significantly in desirable locations and in urban versus rural locations. The positioning of nuclear facility sites relative to social and cultural infrastructures could have a direct impact on the employees' cost of living and their access to infrastructure such as available housing and public education.

8.2. PAY PROGRESSION

Individual performance, competence and skills are common factors that contribute to employees moving up salary bands or ranges. A hybrid approach, where movement is assessed based on several factors, is typical. Such a progression can involve an assessment of the achievements of individual employees based on the competencies required to do their current job, or to move to the next level if the additional competencies are valuable. This type of assessment is important for nuclear organizations that require their employees to gain additional competencies and can be an important motivation, incentive, or retention factor.

8.3. REWARD POLICIES AND PAY AWARDS

Depending on whether an economy is in recession or in growth there are different challenges when it comes to paying salaries and benefits and awarding pay increases. When recession and slow growth are being experienced, organizations may have to resort to being creative in pay management in order to reward their top performing employees. However, when the economy and jobs market is in growth, organizations become more at risk of losing their 'high-performing' staff.

The size of the overall budget for annual pay increases, including performance based increases as well as general pay structure movement (annual pay award or cost of living adjustment), will be influenced by the following factors:

- Ability to pay;
- Inflation;
- Market rate changes;
- The ability to meet legislative requirements (e.g. equal pay, national minimum wage);
- Awareness of regulatory and/or corporate governance standards for executive pay.

HR has a key role to play in managing and supporting the pay review process, consistent with the broader organizational requirements and reward strategy. Procedures need be developed to ensure that pay awards can operate in a timely and effective manner. Annex III includes a list of considerations to facilitate the process.

The reward policy should also link pay increases, promotions and other recognition with safety performance and safety culture requirements. Some organizations publicize these links as a way of demonstrating their commitment to safety, which in turn reinforces the organization's expectations regarding safety culture and performance.

8.4. ROLE OF EMPLOYEE BENEFITS

Employee benefits need to support the retention of employees and encourage them to contribute to achieving the organization's goals and objectives. Before organizations introduce, or revise, benefit provisions, it is important to consider the following questions:

- Why is the benefit needed? What will it achieve?
- How does the benefit support the organization's goals?
- How does the benefit reward the required values and behaviours?
- How does the benefit fit into the HR and reward strategy?
- Does the benefit support the people management and development practices that the organization needs to be successful?
- How will the organization explain how the benefit works and what staff need to do?
- How will the benefit be communicated on an ongoing basis to existing and potential employees?
- What factors will be used to assess whether the benefit is successful in supporting the organization's goal?
- What measures and targets will be used on an ongoing basis to review the benefit's application?
- What is the individual cost to the organization of providing financial benefits?

It is particularly important that nuclear industry employees have benefits and incentives that encourage conservative decision making and early reporting of precursors to potentially serious events. Questioning attitudes that are the integral part of safety and security culture need to be cultivated.

While pay and other financial benefits are important, and getting them wrong can have adverse consequences, they are not the only rewards that employers need to consider. Research shows that non-financial rewards, particularly those around recognition of good performance, can be just as important.

Table 4 contains examples of financial and non-financial benefits that can form part of an employee's overall benefits package.

In flexible benefit schemes, the dividing line between pay and benefits is less rigid than in standard reward packages. This allows employees to vary their pay and benefits package according to their personal requirements. In most schemes, employees are able either to retain their existing salary while varying the mix of various benefits they receive or adjust their salary up or down by taking fewer or more benefits.

Voluntary benefits are where organizations arrange for the purchase of goods and services, often at a discount, by employees. They have no cost to organizations beyond set-up and administrative costs.

TABLE 4. EXAMPLES OF FINANCIAL AND NON-FINANCIAL BENEFITS

Financial	Non-financial
Health insurance	Mentoring — by a person with more knowledge and/or experience for guidance
Bonus payments	Coaching — supported by a coach and given training and guidance
Travel allowance	Career breaks (sabbaticals)
Pension provisions	Flexible working (may not be possible for many roles)
Relocation allowances	Four-day week
Car allowance	Training and development
Share options	Education
Subsidized housing	Career development

8.5. BENCHMARKING

One of the ways to help retain employees is to ensure that remuneration and benefit packages for the current workforce (as well as those being offered to potential new recruits) are competitive within the marketplace to ensure nuclear organizations attract and retain the best people for the job. It is therefore extremely important to undertake regular benchmarking of salaries and benefits.

When undertaking any benchmarking exercise, whether for one role or several, it is important to know that the data being used are accurate, up to date and contain enough information on similar roles in relevant organizations to be a reliable match. The characteristics of the relevant competitor organizations are important, and it is worth considering the geographic/travel-to-work area and the size and types of organizations that are similar to the nuclear industry.

8.6. RETENTION STRATEGIES

The first steps when developing an employee retention strategy are to establish why employees are leaving and then understand the impact that employee turnover has on the organization's performance and its ability to achieve its strategic goals, including the associated costs. This data can be used to develop a costed retention strategy that focuses on the particular issues and causes of turnover specific to the organization.

Depending on the size of the organization, an appreciation of the levels of turnover across different roles, locations and particular groups of employees (such as identified high performers) can also help inform a comprehensive retention strategy and help inform future resourcing decisions.

Tools such as confidential exit surveys and employee engagement surveys can help managers understand why people leave the organization and enable appropriate action to be taken to address it. Ensuring that new employees have realistic expectations of their job and receive sufficient induction training will help to minimize the number of people leaving the organization within the first six months of employment.

Some organizations have found that hiring employees who have ties to the local area can improve retention, particularly at remote locations. To this end, recruiting locally, where possible, even if this means additional training of a lesser-skilled local workforce could be beneficial. This is also an important factor when organizations are considering partnering with education institutions as part of their recruitment pipeline strategy. Partnering with institutions close to nuclear facilities could aid in retention.

Examples of retention activities include offering more attractive pay and benefits and improving induction processes. As well as basic pay and benefits, nuclear organizations need to consider the following elements, all of which have been shown to play a positive role in improving retention and are likely to impact the working environment:

- Job previews: Give prospective employees a realistic job preview at the recruitment stage — do not oversell the job or minimize aspects of the role.
- Career development and progression: Maximize opportunities for employees to develop skills and move on in their careers. Understand and manage people's career expectations. Where promotions are not feasible, look for sideways moves and project opportunities that vary experience and make the work more interesting.
- Employee consultation: Ensure that employees have a 'voice' through consultative bodies, regular appraisals, employee engagement surveys and grievance systems. Where there is no opportunity to voice dissatisfaction and influence outcomes, resigning may be the only option.
- Flexibility: Wherever possible, accommodate individual preferences on working hours and times. As part of this, it is also important to monitor workload and ensure it is manageable within working hours.
- Presenteeism: Be aware of a culture of 'presenteeism' where people feel obliged to work longer hours than are necessary to impress management.
- Fair treatment: A perception of unfairness, whatever the management view of the issue, is a major cause of voluntary resignations. For example, perceived unfairness in the distribution of rewards is very likely to lead to resignations.

9. PERFORMANCE MANAGEMENT

It is important to recognize that not all performance improvement can be managed at the individual level. Three levels of performance improvement can be considered:

- Organizational level: strategy, organizational design/structure and deployment of resources;
- Process level: processes such as operation of the nuclear facility, waste management, ageing management, training and qualification of personnel;
- Job level: job tasks.

The significance of these levels is discussed in Ref. [5]. It is clear that there are interdependencies among these three levels. For example, a lack of strategic planning to develop the competence needed in a new environment can result in inadequate job level performance. Some examples of good practices for achieving effective performance improvement at organizational, process and job levels as well as individual values and behaviours needed for achieving reliable performance (based upon Refs [6, 7]) are included in Annex IV.

The impact of organizational change and process improvement on HRM is addressed in Section 12. The remainder of this section focuses on managing the performance of individuals.

The performance management process aims to maintain and improve employee performance in line with an organization's objectives. It is strategic as well as operational, as its objective is to ensure that employees contribute positively to current organizational objectives and are developed in a way that supports future plans. Ideally, performance will be managed through a range of HR activities. It brings together many principles that enable good people management practices, including learning and development, performance measurement and feedback, and organizational development.

Broadly, managing performance is about setting employee objectives and then providing regular, effective feedback on progress towards them. The objectives need to relate to the organization's goals but also reinforce the desired attitudes and behaviours.

Improving a nuclear organization can include the following attitudes and behaviours:

— The will to communicate problems and opportunities to improve;
— A learning culture and the avoidance of making mistakes;
— Intolerance of 'error traps' that place people and the facility at risk;
— Vigilance and situational awareness;
— Rigorous use of error prevention techniques;
— A will and desire by everyone to 'do their best' every day.

9.1. PERFORMANCE MANAGEMENT PROCESS

At the centre of managing performance are open and honest, yet supportive, conversations about performance that include ongoing feedback. Performance management has the following aims:

— Aligns strategically with the organization's long term goals;

— Establishes objectives through which employees and teams can see how their activities support the organization's mission and strategy;
— Improves performance among employees, teams and organizations;
— Holds employees to account for their performance by linking it to reward and career progression;
— Helps employees develop through structured discussions and opportunities relating to performance, training and development;
— Identifies other learning opportunities (e.g. short term secondments);
— Identifies employees with high potential;
— Integrates with various aspects of the HR strategy.

Managing performance relies on both both formal processes and the routine actions of managers and employees. The formal process is about determining the desired performance around a set of agreed and documented objectives (plan), continually monitoring performance against those objectives (act), discovering and analysing performance gaps, designing and developing effective interventions, implementing these interventions (track) and continually evaluating the results (review) to ensure that the improvement process takes place. This formal process is illustrated in Fig. 3. These are often discussed in meetings between the line manager and employees, known as performance reviews or appraisals. However, managing performance is also about establishing a culture in which individuals and groups take responsibility for the continuous improvement of the organization's processes and their own skills, knowledge, attitudes and contributions.

Managing performance integrates various other HR activities and processes, for example, succession planning, training and development, etc. It needs to be seen as a positive activity with a strong focus on development and not be seen as 'policing' performance.

Each nuclear organization needs to develop formal processes and practices that are relevant to their specific organizational context and their actual (or desired) culture. There also needs to be flexibility within the system itself to account for the different ways teams or functions operate within the organization. However, while performance management will be established as a formal process, it also needs to be considered an integral part of line managers' daily responsibilities.

Line managers and employees can then develop plans and monitor performance regularly. Feedback needs to be given regularly and can be supported by formal performance reviews at agreed points over the course of the year. The plans can also highlight organization-wide processes that are required to support performance, for example, leadership, training, communications and others.

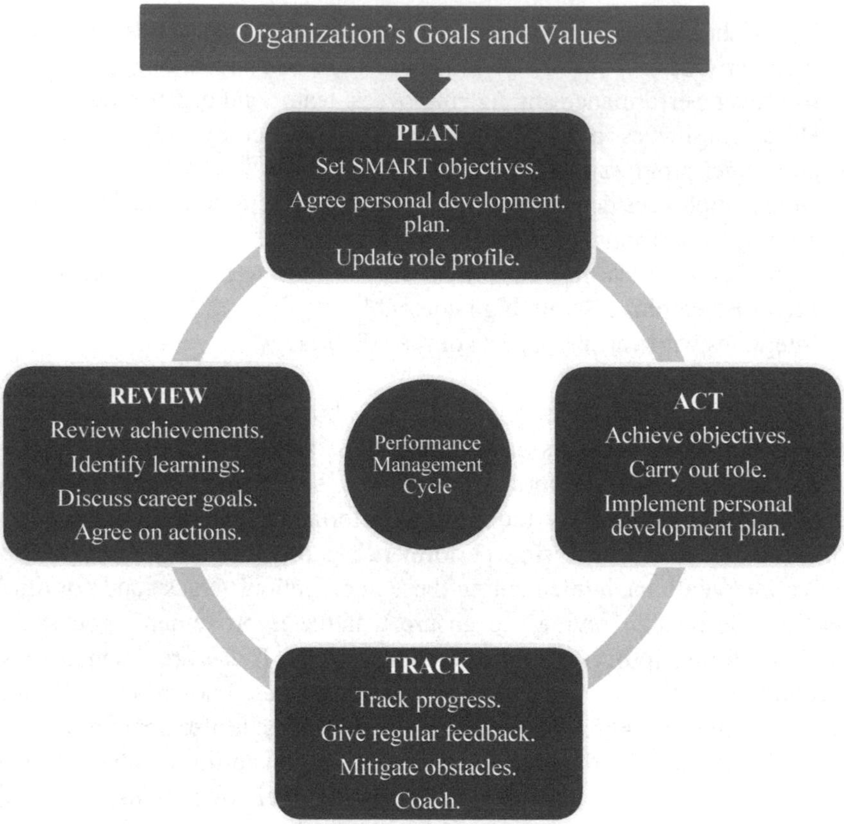

Organization's Goals and Values

PLAN
Set SMART objectives.
Agree personal development.
plan.
Update role profile.

REVIEW
Review achievements.
Identify learnings.
Discuss career goals.
Agree on actions.

Performance Management Cycle

ACT
Achieve objectives.
Carry out role.
Implement personal development plan.

TRACK
Track progress.
Give regular feedback.
Mitigate obstacles.
Coach.

FIG. 3. Performance management cycle.

9.1.1. Setting objectives and performance standards

Organizational strategic goals will provide the starting point for facility and departmental goals, followed by agreement on team and individual goals, performance requirements and development priorities, linked to career aspirations but consistent with the organization's needs.

Setting performance objectives for employees, teams and the organization is an important aspect of managing performance. These objectives can be expressed as targets to be met, ad hoc tasks to be completed by specified dates or ongoing standards to be met. They may be directly related to team or organizational key performance indicators or personal performance indicators; for example, taking the form of developmental objectives for employees. These can be balanced with learning and development objectives and assessments of employees' behaviour, such as how supportive they are of colleagues. For example, social cohesion can be an important factor in driving performance improvement in knowledge organizations, so it is important for such employers to promote teamwork and collaborative behaviour. Whatever their nature, objectives need to be clearly relevant to the overall purpose of the job, team and organization.

9.1.2. Performance appraisal process

Performance appraisal is a process by which a line manager assesses an employee's performance. It is often seen as an annual process, but this need not be the case and regular performance reviews, even if informal, can be helpful to both the line manager and the employee. Assessing and giving feedback on performance is a critical factor in making objectives effective, as monitoring progress towards objectives can be strongly motivational. Therefore, performance appraisals need to be conducted regularly, for example, at the end of a significant task or every few months, based on the nature of the work. They can involve face-to-face conversations between managers and their staff, 360-degree feedback and assessments against performance targets. Performance ratings can be used for administrative purposes (e.g., to inform pay decisions) or to support people development based on managers' judgement of achievement of objectives. Employees' reactions to feedback, and the opportunity to give feedback themselves, are vital factors and are influenced by an individual's personality (e.g., their self-esteem and openness) and how they perceive the appraisal (e.g. as fair and involving).

If the ultimate aim is to improve performance, employee development needs to be emphasized. Performance conversations will also allow employees to learn from their experiences and identify further relevant learning and development opportunities. Personal development plans (PDPs) are often used to set out proposed actions. However, in this respect, it is important to ensure that learning and development opportunities are aligned to the needs of the business as well as the employee.

One output from the appraisal process is the identification of employees with high potential, which becomes an important input to the organization's talent and succession management processes as described in Section 10.

9.1.3. Leadership role in performance improvement

Line managers are instrumental in managing performance and performance improvement. Ideally, they reinforce the links between organizational and individual objectives and give feedback that motivates employees, helps them to improve and holds them to account. Managers need to be suitably skilled and supported by processes that are fit for purpose. However, much of how performance is discussed is shaped by cultural norms. Senior leaders will set the precedent, and line management relationships will in turn shape how employees discuss performance more widely. The desired organizational values to be advocated by all managers needs to include fostering a culture that values prevention of undesirable events, strengthening defences to prevent or mitigate errors and creating an environment that encourages learning and continuous improvement. The desired behaviours of a leader include the following characteristics:

— Eliminating organizational weaknesses;
— Facilitating open communication;
— Demonstrating and reinforcing desired behaviour;
— Valuing error prevention;
— Promoting teamwork.

Managers and employees need to recognize a simple but important fact; effective performance management leads to significantly lower rates and consequences of undesirable events.

9.2. REINFORCEMENT OF DESIRED PERFORMANCE

A well-designed and implemented performance management programme can be a very effective tool for assisting performance improvement. Some form of reward is necessary to reinforce performance improvement. However, it also needs to be recognized that such incentives can be counterproductive if not effectively implemented. For example, incentives that single out individuals for benefits based upon their individual performance can have a negative effect on the motivation of colleagues, particularly if these colleagues are convinced that this individual has received the award unjustifiably or at their expense. One useful approach in this regard is to have incentives based on teamwork performance, rather than individual performance. This approach can be applied equally well to managers and to other personnel. It is important that incentives are carried out not only to correct behaviour in the short term, but also to achieve changes in the long term.

As another example, while managers' objectives need to provide positive consequences for good work and negative consequences for poor work, their actions do not always support this objective. When managers give a top performer more work to do because that individual did it well, this is an example of a commonly occurring negative consequence for individual positive performance. This can be perceived by the employee as being punished for doing good work.

The reverse can also happen when a manager takes away work from a person who works at a low performance level, so rewarding poor performance.

10. TALENT AND SUCCESSION MANAGEMENT

Talent management seeks to identify, attract, develop, engage, retain and deploy individuals who are considered to have either 'high potential' for the future or because they can fulfil critical roles. Managing talent strategically can support organizations to build a high-performance workplace, encourage a learning organization and encourage self-development. It is important that organizations not only respond to current talent challenges but actively anticipate future challenges and opportunities and develop pipelines of talent for the future.

Talent management is driven by the evolving needs of the organization and by a mix of external and internal factors, such as increasingly competitive global markets, nuclear skills shortages and demographic trends.

Succession management is the process of identifying and preparing suitable employees through mentoring, training and job rotation, to replace key personnel within an organization either as their employment terms come to an end, if they leave the organization unexpectedly, or to fulfil future roles.

This section covers the following elements of managing talent and succession:

— The key elements of developing a talent management strategy;
— The key elements of the talent management process, including alignment with other HR processes, future development opportunities and tracking and evaluating talent;
— Roles and responsibilities of key stakeholders;
— Succession management;
— Knowledge management.

10.1. DEVELOPING A TALENT MANAGEMENT STRATEGY

Due to the extensive training and development requirements for many key positions and the associated challenges with external recruitment, the need for nuclear organizations to develop their internal talent is a high priority.

Regardless of which approach an organization adopts, fairness, objectivity, consistency and diversity and inclusion considerations must be applied in all talent management processes. By combining approaches to talent management programmes with diversity policies and activities, an organization can maximize the benefits of accessing and developing talent from the widest possible pool.

There are different models for developing a talent strategy and Fig. 4 gives an example of an integrated approach. Whichever model is adopted, it needs to be aligned with the HRM strategy and organizational culture. The alignment to workforce planning to meet current and future organizational needs is particularly important, including identifying skills needed for key or specialist resources to support the business and people strategy.

The model for a talent strategy in Fig. 4 has four main elements:

- Talent capability: What skills are needed? What are the drivers and priorities? What training and development activities are required?
- Talent pools: How is talent identified? What are the succession planning, talent governance and talent programmes?
- Talent mobility: What are the planning moves, expatriates and career pathways?
- Talent culture: Who has the ownership? How is it aligned to the organizational culture? How is it linked to the diversity and inclusion strategies, performance management, coaching and mentoring? How open and transparent is the organizational culture?

The important point is that all four areas are linked and support each other; for example, it is no good trying to improve the culture if there is not a process or a pipeline.

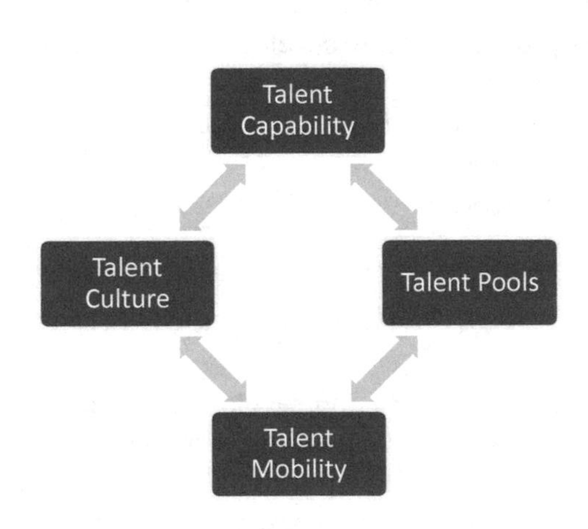

FIG. 4. An example of a model for developing an integrated approach to a talent strategy.

10.2. DEVELOPING A TALENT MANAGEMENT PROCESS

The main aim of a talent management process is to recruit, select and retain people with high levels of competence. There are a number of different formats for developing a talent management process, but many of them include the following key elements.

10.2.1. Alignment with other HR processes

The talent management process needs to be aligned with the workforce planning, recruitment and performance management processes.

The link between talent management and succession management is fundamental whereby succession management is the process which formalizes the identification of talent and facilitates the broader discussion regarding an employee's next career move and any development required to support this. This allows for formal discussions between managers and employees rather than managers identifying training and development opportunities, which may not align with an employee's plans.

Talent management is also closely aligned with training, learning and development to enable talent to benefit from learning and development initiatives including both formal and informal learning interventions.

10.2.2. Development opportunities

Investment in management and leadership development will positively impact on talent retention. The process of succession management, in particular, helps organizations in identifying and preparing potential leaders or technical experts to fill key positions. The following list has a number of options that can be used to help facilitate development opportunities:

- Secondments or lateral moves;
- Job rotation, including rotating assignments as instructors/trainers;
- Deputy/assistant positions, particularly for managers;
- Participation in peer review teams, including external assessments and benchmarking visits;
- Participation in short courses or workshops;
- Assignments to project teams;
- Assignment of a coach or mentor;
- Secondments or training in supplier facilities.

10.2.3. Tracking and evaluating talent management

While there is a need to develop processes to track the performance and progress of talent within an organization, the evaluation of talent management overall is difficult. It requires both quantitative and qualitative data that is valid, reliable and robust. However, the evaluation is necessary to ensure that the investment is meeting organizational needs. For example, one method of evaluating talent management could involve the collation of employee turnover and retention data for key groups such as senior management post holders or those who have participated in talent programmes. Another method would involve monitoring promotion or secondment moves of those people identified as having ‘high potential’ to ensure this group is being developed and given the right opportunities.

10.3. ROLES AND RESPONSIBILITIES

Careful consideration needs to be paid to involve the right stakeholders in developing the talent management process and associated activities and to ensure that clear roles and responsibilities are defined.

10.3.1. Senior managers and line managers

Visible senior management support is important, and a ‘talent panel’ is one method of ensuring the involvement of senior managers, especially when it has representation from across the organization.

Line managers must take responsibility for managing performance, both informally and through an appraisal system, and for identifying and developing talent in their own areas. They also need to be encouraged to see talent as an organizational resource rather than their own resource.

10.3.2. Human resources

HR has a role in providing support and guidance in the design and development of approaches to talent management. They need to understand the key challenges in attracting, recruiting, developing and retaining talented people to meet immediate and future organizational objectives. It is also important to develop and identify career opportunities for employees and creative strategies for unlocking employee potential.

Talent management is now more open and transparent, with many organizations choosing to use formal talent programmes and implement selection processes for talent pools. HR needs to ensure that the selection criteria are robust and applied consistently and support managers when giving feedback to those employees who are not accepted.

10.3.3. Participants

The use of established talent programmes with structured selection processes serves to increase the perceived value of talent programmes and the motivation of participants to perform. Very often one of the positive outcomes of the programmes is that individuals start to think about their chosen career path.

Once participants have completed talent programmes, there is a need to maintain dialogue, by means such as ongoing networking structures or action learning groups. Successful participants are more likely be selected for future leadership programmes as part of their career development, but all participants are equally encouraged to continue with their self-development.

Participants also need to understand the impact regarding some of the outcomes of the process. For example, being part of a talent management pool can mean more frequent job changes, and/or working in different locations.

10.4. SUCCESSION MANAGEMENT

Succession management involves managers and HR personnel identifying individuals as potential successors for each key position and then identifying development activities to help them prepare for their next position. This is important for nuclear organizations because it can take years to develop people for specialist positions.

One of the key objectives in succession management is to match the organization’s future needs with the aspirations of individuals. Providing development opportunities that offer new challenges and are more promising than those that can be found elsewhere is an effective way to retain talented people.

When enacting succession plans senior managers have to pay attention to the resilience of the team or teams when internal moves and appointments are being considered. The following list includes criteria for consideration:

— Length of time and performance in post;
— Number of years of managerial experience;
— Number of years of technical experience in terms of operational, maintenance, etc.;
— Number of years of experience in terms of their technical background (e.g. scientist, engineer);

— Other relevant generalist experience (e.g. project management, safety case);
— Other relevant technical experience (e.g. structural integrity);
— Impact of person/people moving into new posts versus posts left behind.

The potential loss of people with critical knowledge and skills can be addressed through conducting risk assessments to determine the potential for loss of critical knowledge and to enable nuclear organizations to use this knowledge to improve the competence of new and existing personnel (see Ref. [8]). Critical posts need to be identified and where possible successors also need to be identified as either permanent or as a 'temporary safe pair of hands' with agreed action plans to mitigate risk and engage in external recruitment, if required.

10.5. KNOWLEDGE MANAGEMENT

Knowledge management is acknowledged as being important for nuclear organizations and they need to have a knowledge management programme to ensure that critical knowledge is transferred before people leave or move within an organization. More information on knowledge management can be found in Refs [8–10].

Knowledge management is closely aligned with workforce planning and succession and talent management. When an organization identifies that it will lose key skills (e.g. retirements), sufficient lead time is required to allow for recruitment, training and knowledge transfer. This process needs to take into account the complexity of the role, the skills and training required and the extent of supervisory and management involvement.

11. EMPLOYEE ENGAGEMENT AND WELL-BEING

Employee engagement is a workplace approach intended to ensure that employees are motivated to contribute to organizational success and have an enhanced sense of their own belonging and well-being. Achieving an organization's goals is increasingly dependent on the performance of individual employees, and employee engagement has become a key part of the employment relationship.

Promoting and supporting employee well-being is at the heart of any organization that wants to champion better work and working lives. An effective workplace well-being programme will deliver mutual benefits to people as well as to the economy and wider society. Due to the fast-changing world of work and the fluctuating demands placed on organizations and employees, our understanding of health and well-being needs to evolve constantly to mitigate and optimize the impact on people.

11.1. EMPLOYEE ENGAGEMENT

Organizations want engaged employees because, as well as being happier, healthier and more fulfilled, they are likely to be more motivated and deliver improved performance. Positive relationships are evidenced with profit, revenue growth, productivity, innovation, staff retention, efficiency and health and safety performance. For nuclear organizations, nuclear safety and security are overriding priorities and therefore employee engagement has a key role to play.

There are four fundamentals of an effective employee engagement strategy:

- Leadership that gives a strong strategic narrative about the future direction of the organization supported by a business plan that is communicated and understood;
- Line managers who motivate, empower and support their employees;
- An 'employee voice' throughout the organization, to challenge or reinforce the status quo and involve employees in decision making;
- 'Organizational integrity' that includes stated values which are embedded into the organizational culture.

Successful employee engagement strategies will build on good people management and learning and development practices. They will be holistic and focus not only on employees' work engagement and well-being, but also help employees to see clear links between their work and the organization's mission, vision and values.

Many large organizations conduct regular employee attitude or engagement surveys, often alongside focus groups or other forums to gain employee insight. The benefit of a survey approach is that organizations can get a representative view from employees across the organization, and establish benchmarks, which can be used by subsequent activities and will help to identify trends, both positive and negative. The benefit of qualitative methods, such as focus groups, is to hear the true voice of employees and get a richer, less constrained understanding than from pre-set questions and options.

11.2. EMPLOYEE WELL-BEING

Health and well-being are not an 'add-on' or 'nice to have' activity. All nuclear organizations need to place employee well-being at the centre of their business and view it as a vital source of value creation, delivering overall improved organizational health and performance.

An effective employee well-being strategy needs to go beyond a series of standalone initiatives. There is no 'one-size-fits-all' approach to designing a health and well-being strategy; rather its content needs to be based on the organization's unique needs and characteristics. A well-being strategy also needs to be multidimensional in terms of acknowledging not only the physical and psychological factors that affect health and well-being, but the wider cultural and societal contexts too.

Investing in well-being makes good business sense and can lead to increased resilience, greater innovation and higher productivity. An effective health and well-being programme is likely to include the following elements:

- — Health promotions;
- — A good working environment;
- — Workplace assessments;
- — Flexible working;
- — Positive relationships;
- — Opportunities for career development;
- — A healthy management style.

For well-being initiatives to achieve significant benefits, they need to be integrated throughout an organization, embedded in its culture, leadership and people management activities. The HR profession is in a unique position to understand the needs of both workforce and organization and to deliver the benefits of well-being.

11.2.1. Roles and responsibilities

Having said that a well-being strategy needs to be multidimensional, it follows that there are multiple partners involved in ensuring that the strategy is both appropriate and effective. Their roles and responsibilities need to be clearly defined and communicated.

11.2.1.1. HR professionals and practitioners

HR professionals have a key role in ensuring that senior managers understand the importance and benefits of a health and well-being strategy and that the developed strategy is right for the organization's needs. They also need to provide expert support to managers and employees to ensure effective implementation. Some organizations have dedicated occupational health practitioners who are directly responsible for ensuring workplace health and these will work in partnership with the HR professionals.

11.2.1.2. Senior managers

Senior management commitment to the health and well-being strategy is central to its success. It is their role to ensure an appropriate strategy is developed and integrated into the wider management system. They will also role-model the appropriate activities and behaviours for the benefit of line managers and employees.

11.2.1.3. Line managers

Line managers have a critical role to play as much of the actual implementation of the strategy will be their responsibility day to day. It is the line managers who interact most frequently with employees and have the most potential to influence them and to spot potential problems. To be effective, they will need to have good people skills — in particular, influencing' and conflict and stress management skills. Training in these and other people management skills needs to be included in the organization's management training programmes.

11.2.1.4. Employees

It is important for management to stress a holistic approach to health and well-being and employees need to understand their responsibilities both within the organization's strategy and in their lives outside work as well. Employees need to participate in the initiatives and activities proposed by management and to provide feedback on the effectiveness of those activities.

12. EMPLOYEE TURNOVER AND RETIREMENT

The training and development of nuclear facility employees require large investments of time and money, as well as having a considerable lead time. In the past, the majority of the workforce was expected to remain with the same employer for their entire career. Increasingly, employees have greater mobility to take other positions, whether with similar organizations, in other industries or even in other countries. These factors highlight the importance of managing turnover and monitoring the extent to which employees are satisfied with their jobs and associated benefits.

Employees resign for many different reasons; sometimes it is the attraction of a new job or the prospect of a period outside the nuclear industry that attracts them. On other occasions they leave as a result of dissatisfaction in their present jobs, for example, from a lack of career opportunities, organizational changes or a poor relationship with a line manager leading to disengagement. Of course, they could be leaving because they have reached their natural retirement age.

Leaving an organization (often called exiting) is an extremely important part of the employee life cycle, yet many organizations undervalue this process. Just as induction and orientation is the process by which an organization ensures that a new starter is integrated into the business culture as efficiently as possible, exiting is an opportunity to ensure that an employee leaves the organization in the most effective and efficient manner.

Whatever their reason for leaving, early engagement with these employees is important to be able to properly plan their exit and conduct any necessary knowledge transfer activities. This is easier to predict for employees as their natural retirement age but is especially important in cases where there is no fixed retirement age.

12.1. INVESTIGATING WHY PEOPLE LEAVE

It is important to understand why people leave an organization. However, obtaining accurate information or reasons for leaving can be difficult. Individuals may be reluctant to voice criticism of their managers, colleagues or the organization generally, preferring to give some less contentious reason for their departure.

Where exit interviews are used to ask about the reasons for leaving, the interviewer needs to not be a manager who has responsibility for the individual or who will be involved in future reference writing. Confidentiality needs to be assured, and the purpose of the interview explained. Using someone seen as independent within the organization, or an external provider, to conduct exit interviews will help organizations capture more accurate data about why people are leaving, as individuals are more willing to be truthful when there is reassurance of anonymity. Alternative approaches to collecting exit data involve the use of confidential questionnaires sent to employees on exit or a period of time after their departure.

It is also important to consider the experiences of those still employed within the organization; conducting an exit interview will gather useful data but using information from current employees (such as employee surveys) will help to pinpoint retention issues before they lead to people leaving.

12.2. CONSIDERATIONS WHEN PEOPLE LEAVE

The exiting process needs to be effectively deployed to manage the termination of an employee including collecting any company materials (e.g. passes, keys), outstanding payments and managing any possible outstanding relocation payments owed. Managers also need to be aware of any clauses within the contract of employment relating to confidentiality and intellectual property and understand their meaning and remind the employee if appropriate. A checklist needs to be in place as a reminder to cover all items and to ensure the safe return of all company equipment and documentation. Handover arrangements need to be managed effectively and the knowledge transfer needs to be completed to the satisfaction of all key stakeholders.

Whether people are leaving the organization through resignation or retirement then the knowledge transfer process needs to be managed accordingly. Where people have known retirement dates knowledge transfer can be managed over a longer, more planned timetable. Where people have resigned the normal notice periods may only be three months so knowledge transfer will have to be planned and managed accordingly over a much shorter time frame.

There are a number of other reasons why it is important for nuclear organizations to invest time and effort into their exit process, including those discussed below.

12.2.1. Turnover in the first year of employment

In many organizations, a large proportion of employee turnover consists of people resigning in the first few months of employment. Valuable lessons need to be learned from those leaving. Poor recruitment and selection decisions, both on the part of the employee and employer, can often be to blame, or poorly designed or non-existent induction programmes. The first impressions of an organization can influence an employee's decision to leave a business sooner rather than later. Also, expectations are also often raised too high during the recruitment process, leading people to compete for and subsequently to accept jobs for which they are in reality unsuited.

12.2.2. Data security

Since all nuclear organizations have processes in place to ensure that data is secure and access to internal systems is restricted, it is important to ensure that access to this information is revoked when an employee leaves the organization. Many information management systems will automate these processes and eliminate the risk. In addition to revoking access, the exit process will facilitate the return of physical inventory, such as computers, cellular phones and data storage devices.

12.2.3. Protecting the employer brand and retaining a positive candidate attraction strategy

Including interviews in the exit process allows employees who are leaving the organization to give feedback about their employment, which can be of value both to the individual and the organization.

If an employee has a positive experience with an organization and leaves on good terms, they are more likely to recommend the brand to friends and family. With web sites such as www.glassdoor.com growing in popularity as a reference for potential talent to research an organization, it is important to maximize the chances of receiving positive reviews from both existing and ex-employees.

Social media is being used increasingly to talk about good, and more importantly, bad experiences. It is therefore imperative not to let employees leave feeling that they have to inform others of any bad experiences within the organization. Bad publicity on the scale that is possible through social media can have a damaging impact on an organization over something that often could easily have been avoided.

12.2.4. Return of ex-employees

In some cases, ex-employees may return to work for an organization, especially if their exit was due to voluntary severance or retirement. This is particularly true in the case of the nuclear 'family' where many ex-employees will stay within the industry in some capacity and therefore employees leaving in a positive manner will mean the industry will not miss out on the expertise and talent that they can offer the organization. If an ex-employee returns, they can usually be inducted much sooner than an employee with no prior knowledge of the organization and its culture.

13. CHANGE MANAGEMENT

Organizational change is a constant in many organizations today and can be driven by a number of different factors, including legislative, economic, market-driven and technological. The consequences of not managing this change effectively can be long lasting, so it is important that HR professionals understand the potential issues and equip themselves with techniques to support change. Changes can include organizational structural changes, process improvement, company mergers, or reductions in resources as a result of outsourcing or a need for cost reduction. Regardless of the cause, the effort and resources required will depend on the nature and significance of the change. Failure to manage change effectively may adversely impact employer brand and individual performance and therefore have a significant impact on the overall organization's performance.

13.1. EFFECTIVE CHANGE MANAGEMENT

For nuclear organizations it is important to have a systematic approach to managing change and improvement proposals, with each change following a documented process under an agreed policy to ensure focus is maintained on safe, reliable operation. This can also be required by regulatory bodies and is critical in understanding and controlling the risks. Inappropriately conceived or managed changes to processes, systems, organizations, working practices or resource levels can introduce significant risk to the business. Too often managers are implementing change with inadequate consideration or understanding of the risk they are introducing to the organization.

Change is complex, and there is not an easy single solution to managing it. However, there are key areas of focus for effective change management:

- Identifying the need for change: It is vital to be clear about the benefits expected to be gained from the change.
- Consider all options: Open discussion and debate needs to be encouraged to support ownership of the proposed change and enable a more proactive change process to happen.

— Understanding the context: Managers who are responsible for designing and managing change need to be able to understand the context and consider all the related issues such as impact on structure, systems and processes.
— Align strategy and culture: For high level change to succeed, managers also need to align strategic and cultural aspirations, especially if this requires a shift in behaviours.
— Describe the change: Narratives and stories, including understanding the start and end of the change, can be used as devices to make the content and implications of the change easier to understand, enhancing a manager's ability to translate change into meaningful actions for themselves.
— Consult stakeholders: Develop a stakeholder plan and ensure all key stakeholders have been identified and consulted (e.g. the nuclear regulator and trade unions). When an individual or group is resistant to change, they may engage in acts to block, slow down or disrupt an attempt to introduce change. However, resistance is not necessarily negative and may be a clear signal that the change initiative needs rethinking or reframing. Therefore, it is important to avoid this assumption and instead try to diagnose the cause of employee resistance, as this will help determine how to address the issue. In more democratic workplaces, the actions of stakeholders and employees who raise concerns about change will not be labelled as resistance, but instead reframed and reinterpreted in terms of legitimacy of the 'employee voice'.
— Building trust: High levels of trust will deliver the enabling conditions in which significant change can thrive. Change managers need to emphasize their trustworthiness by demonstrating their competence to design and lead change intelligently and their honesty and integrity in the way they attend to the needs of the organization, employees and the wider community.
— Emotional awareness: Change is often an emotional process and so emotional awareness by those designing and leading change is required to anticipate and plan for reactions.

While change is generally driven and implemented through line managers, HR staff can play a significant role in ensuring the above areas are properly addressed.

13.2. DEVELOPING THE CHANGE PROPOSAL

When developing the change proposal (the way in which change should be addressed) it is important to consider the following aspects:

— Ensure that the roles affected are identified along with systems, processes, interfaces, documents, etc.
— Timescales need to be considered in achieving the desired outcomes and communication plans also included.
— The project team and structure need to be identified, alongside any training requirements. The change needs to be seen as a 'project', and the use of effective project management disciplines is essential for success. Insufficient relevant training in change management and leadership skills can all impact negatively on the effectiveness of any change initiative.
— Risks need to be identified which may prevent the change being achieved, reduce or eliminate any benefit of making the change or have the potential for an unplanned event to occur with economic or safety implications. Latent risks also need to be addressed (i.e. is there anything that can cause problems in the long term rather than immediately).
— There needs to be an enabler to address each risk identified. Enablers are specific planned activities to ensure the change is delivered properly and potential risks are avoided (or at least mitigated), and they must be specified in the planning phase. Any enabler must be SMART (specific, measurable, achievable, realistic and timely). A pre-implementation review may be needed.
— Countermeasures (an activity or measure) that might be needed if a potential risk manifests itself during or after implementation. Therefore, if a change is not going to plan, it might be a case of reverting to the original structure either partly or totally, adding additional resources or reviewing and revising the change.
— Performance measures need to be in place to monitor the change and ensure that nuclear safety is not affected. These measures need to be appropriate and provide direction, if possible. Other measures can include feedback from appraisals, informal discussions, team meetings and feedback from stakeholders.
— A post implementation review needs to be carried out against the original change proposal, usually three or six months after the change has been implemented.

13.3. HR ROLE IN MANAGING CHANGE

Line managers and HR professionals have a significant role to play in any change management process. They often play a critical role behind the scenes working in partnership with senior and line management to help facilitate translation of the overall vision through communication, use of relevant techniques, and changes to well established systems or behaviours.

In some cases, major organizational change or business improvement will require significant reductions in staff, redeployment of staff to other areas of the business or to another company (e.g. outsourcing) or changes in terms of conditions. In consultation with senior management, HR will need to define the terms and conditions that will be applied and the process under which this will take place. They may also need to develop specific training for line managers involved in counselling staff and provide the relevant support services to impacted employees.

HR professionals will need to balance the interests of the organization against the needs of the employees. This can be challenging during a period of significant change. The ability to apply situational judgement and demonstrate moral integrity is what will enable them to be trusted advisors, and help the organization create long term sustainability.

13.4. MANAGERS' ROLE IN MANAGING CHANGE

Managers must be able to introduce and manage change to ensure the objectives of change are met, and that they gain the commitment of employees, both during and after implementation. Often, at the same time, they also must ensure that the focus remains on the safe, reliable operation of the organization.

Managers at all levels need to provide leadership in setting and communicating organizational change, benefits, values and ethics. Because of the high visibility of the nuclear industry with the public and decision makers, clear and open communication regarding any change is particularly important to build trust and respect. Some changes may require prior regulatory approvals and it is the senior managers' role to oversee this process.

13.5. KNOWLEDGE MANAGEMENT AND LEARNING ORGANIZATIONS

It is recognized that the nuclear industry is very much knowledge based and relies heavily on the knowledge of skilled employees. The ageing workforce and the risk of losing accumulated knowledge and experience have drawn attention to the need for better management of nuclear knowledge. The IAEA has a number of publications related to the need for effective strategies and policies for knowledge management.

Organizational learning can be defined as the process of improving performance through better knowledge and understanding. Therefore, openness, shared knowledge and continuous improvement, which are essential values in a nuclear organization to ensure safety and maintain efficiency, will also help to make it a learning organization. If an organization stops searching for improvements and new methods of working, via benchmarking and seeking out good practices, there is a danger its performance will decline relative to others. HR has a key role to play in supporting the organization in adapting to a changing environment through activities such as recruitment, career and knowledge management, communications, mentoring and change management.

A learning organization is able to tap into the ideas and concerns of those at all levels of the organization and enhancements in safety can then be sustained by ensuring that the benefits obtained from improvements are widely recognized by individuals and teams. Nuclear organizations are required to manage safety as a major component of their activities and must learn from precursors and near misses, which is why processes exist to share worldwide experience. Therefore, the emphasis needs to be on 'getting it right first time', encouraging a 'questioning attitude' and avoiding latent errors by learning from events and good practices.

14. SAFETY AND ORGANIZATIONAL CULTURE

The nuclear industry is different to many other sectors due to the potentially significant impacts of any errors that occur. This necessitates greater levels of regulatory oversight, heightened safety and security arrangements, more in-depth training, checking and contingency planning, strict adherence to procedures and a focus on individual employee behaviours. Human behaviour is recognized as an important aspect in ensuring the prevention of mistakes, incidents and accidents. The way people behave in an organization is very much influenced by the culture of the organization.

Hence, it is essential that the organizational culture is one where high standards of personal and professional behaviour are set, modelled and supported by senior and line managers. This will create the right environment for the creation of a strong safety culture. HR staff have an important role to play in supporting safety culture by ensuring that all HR activities, especially training and development activities, reflect and promote the correct behaviours. Training programmes on safety culture and personal behaviour will also be important supporting elements.

Given the increasingly global nature of the nuclear industry, cultural sensitivity is essential when establishing or developing the organization's culture. It is very likely that personnel, whether staff or contractors, will have different business practices, communication and management styles. Understanding the culture of employees within the nuclear industry is especially important when undertaking day to day activities and tasks and looking to accomplish objectives. When moving into a new organization people will need to understand the following aspects of the organization:

- How employees communicate;
- How employees perceive spoken and written communications;
- How employees view punctuality, timescales and deadlines;
- How likely employees are to ask questions or highlight problems;
- How employees respond to management and authority;
- How employees make decisions.

14.1. FOSTERING A JUST AND OPEN CULTURE

For nuclear organizations there must be an environment that recognizes the human potential for error and clearly defines acceptable behaviour in a consistent manner. Where this culture exists, it represents a 'just culture', meaning it treats people fairly. An organization that has a just culture has the following attributes:

- Recognition of fairness related to the identification and resolution of personnel performance problems;
- Distinction between honest mistakes and intentional shortcuts with respect to discipline;
- Free flow of information across and between all levels of an organization (better known as openness and trust);
- High levels of self-reporting.

It is important to promote a working environment and culture in which people are encouraged to openly report mistakes that they themselves made with the knowledge that they will not face any disciplinary or punitive action for doing so, providing they are genuine errors. They understand that their openness will be used solely for learning and preventing others from making a similar mistake. The benefit of creating such an organizational environment is that everyone will act to avoid error and prevent its damaging consequences, whether they are safety, financial, reputational or, as is often the case, all of these.

Line managers have a powerful role to play in communicating that they want people to openly report actual and potential errors and that they will be treated as unintended and people will not be disciplined for doing their best, even if this is what led to an event. Engaging line managers is about ensuring they understand how the organization can protect itself from harmful events through proactive reporting and how this is only possible with an open culture in place. Their roles are to support this, to recognize that there is a process in place to deal with errors and to let the system work.

14.2. WHISTLEBLOWING

Nuclear organizations will always try to generate an open culture that supports the reporting of incidents. However, there may be certain circumstances where a 'whistleblowing' policy can be used which allows employees to report suspected misconduct or certain types of wrongdoing on a confidential basis. This type of policy will usually contain an external confidential reporting process that provides a safe and secure means by which employees can speak up with confidence. This type of policy will usually be linked to an organization's code of conduct policy. The HR function has an important role to play in ensuring the integrity of this process and the protection of individuals who use it. Some organizations adopt an anonymous reporting system to allow employees to raise their concerns.

15. STAKEHOLDER ENGAGEMENT

The purpose of stakeholder engagement is to enable stakeholders to make their views known and to work together to ensure that these views are addressed/considered. At the same time, it needs to be recognized that the aim of an effective stakeholder involvement programme is not to gain 100% agreement, but for stakeholders to understand the basis for a decision and thus have greater trust that the decision was appropriate. It needs to also support the development of consensus, where possible.

A stakeholder is a person, company or group with a concern or interest in ensuring the success of an organization, business, system or entity. However, there are also some stakeholders, such as campaign groups, who are indifferent about the success of the organization and who care only about safety or the environment, for example. Some of the most important aspects of stakeholder involvement are to find out the real concerns of the stakeholders, act upon them and treat them with respect. This includes the understanding of minor issues and preventing these issues from escalating into a situation that erodes stakeholder confidence. Figure 5 below shows some examples of the types of stakeholders in the nuclear field.

Nuclear facilities in Member States often interact with local, regional and national entities either supporting or opposing the construction, operation or early decommissioning of any given facility (Fig. 5). Therefore, emphasis needs to be placed on engendering trust in the community (local or national) of the organizations and institutions involved in the process. Reliability, responsibility and fairness are attributes that foster trust in those participants in decision making processes.

Given the timescales involved in developing, constructing, operating and decommissioning nuclear facilities, which in the case of a new programme can be about 100 years [11], obtaining and maintaining stakeholder support is important. It is therefore vital that engagement with younger generations forms an important part of any stakeholder involvement process, given that its members will be impacted throughout their lives and are the decision makers of the future.

Having identified concerns and sensitivities among the various stakeholder groups and how those groups can impact the programme or facility development in question, there is then a need, within decision making processes, to clearly assign responsibilities and roles for stakeholder involvement in these processes. This needs to include explanations of what decisions are required and how stakeholders can influence them, or if not, why not. Typically, responsibilities for involvement/engagement with employees (internal stakeholders) and other personnel who work onsite are shared by the HR function and line managers.

Stakeholders in the Field of Nuclear Energy

- Employees
- Contractors
- Public
- Government
- Non-governmental organizations
- Media
- Labour unions
- Trade unions
- Educational institutions

FIG. 5. Examples of stakeholders in the field of nuclear energy.

Even if external stakeholder engagement is managed by another group, it is important that the HR function works closely with them, to ensure consistency between internal and external stakeholder relations.

The following steps are common to implementing stakeholder involvement programmes for any nuclear facility or programme:

- Develop a strategy for stakeholder involvement;
- Develop plans for implementing this strategy;
- Ensure that the capacity to effectively implement these plans is available;
- Implement these plans;
- Monitor continually the effectiveness of these actions and look for ways to improve.

Internal and external communications are equally important. Effective internal communication can help to build a team that clearly understands the different yet equally demanding roles of experts (remembering also that all industry staff are potentially 'spokespersons' for the industry and so need to be well informed, even 'trained', themselves). Effective external communication can present the expertise of the organization to broad external audiences.

15.1. CORPORATE AND SOCIAL RESPONSIBILITIES STRATEGY

Most successful organizations put corporate and social responsibilities (CSR) at the heart of their organization. For nuclear organizations this has to be a key priority in making them an attractive employer for the following reasons:

- The need to work in partnership with local communities;
- The protection and promotion of brand awareness;
- The need to build trust with employees and other key stakeholders;
- The protection and sustainability of the environment.

The role of HR becomes easier when there is a successful CSR agenda which supports such activities as recruitment, retention and talent management, therefore supporting workforce planning for the longer term.

15.2. WORKING IN PARTNERSHIP

Working in partnership with key stakeholders such as suppliers, contractors and trade unions helps to recognize respective roles and responsibilities, establish shared values and ensure a common purpose and objectives. Given the long time frames associated with the nuclear industry, working in partnerships contributes to stability and behaviours that are required in dealing with each other.

HR has the responsibility for managing the partnerships with trade unions or workers forums on the basis that a positive and inclusive approach to employee relations is conducive to the achievement of business objectives.

16. DIVERSITY AND INCLUSION

Due to the nature of the nuclear industry, far more collaboration exists between companies that would normally call themselves competitors than would be typical in other industries. International cooperation and partnerships bring opportunities to draw on a broad pool of resources without which an operating organization can struggle to maintain complete capability. International cooperation in science and development can help to optimize the deployment of scarce manpower and the construction and operation of large scale research and engineering test facilities. Many companies are now operating on a global basis and develop specialist teams that provide services to organizations in many Member States.

As stated previously, the nuclear workforce has become more mobile, which has led to different people and cultures working together. Promoting and supporting diversity in the workplace is, therefore, an important aspect of good people management and ensures that everyone in the organization is valued as an individual.

To realize the benefits of a diverse workforce it is vital to have an inclusive environment where everyone feels able to participate and achieve their potential. While legislation varies in different countries and sets minimum standards, an effective diversity and inclusion strategy goes beyond legal compliance and seeks to add value to an organization, contributing to employee well-being and engagement.

It is important to recognize that a 'one-size-fits all' approach to managing people does not achieve fairness and equality of opportunity for everyone. A multicultural workforce means there are different personal needs, values and beliefs. Good people management practice demands that peoples' working conditions are both consistently fair but also flexible and inclusive in ways that are designed to support both individual and organizational needs.

16.1. DEFINING DIVERSITY AND INCLUSION

Diversity is where difference is recognized and emphasized, but not actively leveraged to drive organizational success. Inclusion is where difference is seen as a benefit to make use of, and where perspectives and differences are shared, leading to better decision making. In an inclusive working environment, every person feels valued and that their contribution matters, and they are able to perform to their potential, no matter their background, identity or circumstances. An inclusive environment enables a diverse range of people to work together effectively, irrespective of their age, race, gender or disability, which is essential within the nuclear industry. For nuclear organizations that operate internationally, as many do, their approach to managing diversity will need to take account of the ways that individual working styles and personal preferences are influenced by the social environment and culture that people have become accustomed to. There is acknowledgement of the benefit of having a range of perspectives in decision making and the workforce being representative of the organization's customers and stakeholders.

The main advantages and importance of considering diversity and inclusion in their broadest sense is as follows:

— People want to work for organizations with good employment practices and they also want to feel valued at work.
— Diversity and inclusion strategies help to develop an open, inclusive and 'blame-free' working culture.
— Diversity and inclusion strategies are important in recruiting and retaining the skills and talent required for an organization's long term sustainability.
— Organizations that consider their CSR strategy in the context of diversity can avoid social exclusion and low economic activity rates, which can limit recruitment and talent pools.
— Communicating and adhering to a strong set of values and ethics sends an important message to the public, suppliers, stakeholders and employees.
— When incorporated into a strong employer brand, diversity and inclusion can help to attract and retain talent.
— Different perspectives lead to innovation and different approaches to problem solving, and good leaders will recognize the value of seeking out and considering different opinions.

16.2. BUILDING A DIVERSITY AND INCLUSION STRATEGY

There will be different challenges for different organizations depending on what stage of their organizational journey they are on. When building a diversity and inclusion strategy, it is advisable to observe other organizations and learn from them as well as seeking out best practice. It is recognized that an organization's diversity and inclusion strategy will need to be consistent with national norms, requirements and practices, but also needs to incorporate international good practices.

Once the diversity and inclusion strategy has been developed and implemented, it needs to be routinely evaluated and reviewed to ensure it is aligned with the organization's goals and considers current best practices regarding diversity and inclusion; then, any necessary corrective actions should be taken.

REFERENCES

[1] MINCHINGTON, B., Your employer brand: Attract, engage, retain, Employee Value Proposition (2006), https://catalogue.nla.gov.au/Record/3799989?lookfor=author:%22Minchington,%20 Brett%22&offset=1&max=1

[2] INTERNATIONAL ATOMIC ENERGY AGENCY, Assessing Behavioural Competencies of Employees in Nuclear Facilities, IAEA-TECDOC-1917, IAEA, Vienna (2020).

[3] INTERNATIONAL ATOMIC ENERGY AGENCY, Nuclear Power Plant Personnel Training and Its Evaluation, Technical Reports Series No.380, IAEA, Vienna (1996).

[4] INTERNATIONAL ATOMIC ENERGY AGENCY, Authorization of Nuclear Power Plant Control Room Personnel: Methods and Practices with Emphasis on the Use of Simulators, IAEA-TECDOC-1502, IAEA, Vienna (2006).

[5] RUMMLER, G., BRACHE, A., Improving Performance: How to Manage the White Space in the Organization Chart, Jossey Bass Business and Management Series, Jossey-Bass, San Francisco, CA (1990).

[6] INSTITUTE OF NUCLEAR POWER OPERATIONS, Human Performance Fundamentals Course Reference, National Academy for Nuclear Training, Revision 6, INPO, Atlanta (2002).

[7] INSTITUTE OF NUCLEAR POWER OPERATIONS, Performance Objectives and Criteria, Rep. 05-003, INPO, Atlanta, GA (2005).

[8] INTERNATIONAL ATOMIC ENERGY AGENCY, Risk Management of Knowledge Loss in Nuclear Industry Organizations, IAEA, Vienna (2006).

[9] INTERNATIONAL ATOMIC ENERGY AGENCY, Knowledge Management for Nuclear Industry Operating Organizations, IAEA-TECDOC-1510, IAEA, Vienna (2006).

[10] INTERNATIONAL ATOMIC ENERGY AGENCY, The Nuclear Power Industry's Ageing Workforce: Transfer of Knowledge to the Next Generation, IAEA-TECDOC-1399, IAEA, Vienna (2004).

[11] INTERNATIONAL NUCLEAR SAFETY ADVISORY GROUP, Safety Culture, INSAG-4, IAEA, Vienna (1991).

Annex I

EXAMPLE TEMPLATE OF A JOB ANALYSIS FORM

This annex provides a template that can serve as the basis for the development of a job analysis form.

HIRING MANAGER	Name				
	Job title				
	Contact details				
DIVISION/TEAM/DEPARTMENT					
POSITION JOB TITLE					
NUMBER OF POSITIONS					
ESTABLISH THE NEED/URGENCY					
— How has this need arisen?					☐
— Is it temporary or a permanent role? What is the justification for temporary or permanent?					☐
— Start date? (What implication is there of delay?)					☐
— How long have you been looking? (If a long time, why?)					☐
— Is it internal work or client billable work?					☐
— Interviews happened or planned? Any offers made?					☐
— Have the CVs been received (quantity and quality)?					☐
JOB AND PERSON SPECIFICATION					
Is there a formal job specification?		Yes	☐	No	☐

ROLE DESCRIPTION	
— Describe the exact deliverables in this role/project.	☐
— What exactly will the responsibilities of this person be?	☐
— Who are the stakeholders this person will need to interact with (internal and external)?	☐
— What is the work stream or project timelines?	☐
— Which locations will they work in and what travel will be necessary?	☐
— How will their success/performance be measured?	☐
— Examples of tasks/situations they will face in an average day/week?	☐
ESSENTIAL	
— In priority order, what are the essential skills/experience this person will need to have?	☐
— What are the needed specific project experience, technologies, environments, languages, management, disciplines, regulatory systems, countries/cultures, qualifications?	☐
— How long should a person have had each experience or how long should a person have used each skill?	☐
DESIRED	
— In addition to the essential skills, what other skills/experience would be desired?	☐
— Think of someone you already have in your team/company with similar experience and describe specifically what it is you like about their skills/ experience/background/working style.	☐
— When reviewing a CV, what are the specific things that you look for, in order of priority (e.g. type of company, working locations, job titles, length of experience, gaps, time at each employer, education)?	☐

HIRING PROCESS				
What is the process that will happen from this conversation to the employee's starting day at work?				
— Agree when CVs will be sent and to whom.				☐
— Who is involved in CV review/interview/decision process?				☐
— Agree exact time and date for CV feedback.				☐
— Agree interview slots — time, date and who is involved.				☐
— How long to go from interview to offer and who is involved?				☐
— Once an offer is accepted, what else needs to be done before they start? (For example, reference checks, security clearance.)				☐
— What is the total length of the hiring process from now?				☐
— What factors could slow the hiring process down? How can these obstacles be avoided?				☐
SALARY/PAY RATE				
— Is the salary approved? How much? (Hourly/daily/monthly/annual.) What flexibility could there be (and under what circumstances)?				☐
— Hours per week? (Cap? Flexible hours? Four-day week?)				☐
— Expenses/per diems payable?				☐
FULL JOB SPECIFICATION	Yes	☐	No	☐
INTERVIEW SLOTS	Yes	☐	No	☐
START DATE (PROVISIONAL)	Yes	☐	No	☐
BUDGET APPROVED	Yes	☐	No	☐
Agree exact time/date of next contact and actions for each party.				

<table>
<tr><td>HIRING MANAGER (cont.)</td><td>Name</td><td colspan="4"></td></tr>
<tr><td></td><td>Job title</td><td colspan="4"></td></tr>
<tr><td></td><td>Contact details</td><td colspan="4"></td></tr>
<tr><td colspan="2">DIVISION/TEAM/DEPARTMENT</td><td colspan="4"></td></tr>
<tr><td colspan="2">POSITION JOB TITLE</td><td colspan="4"></td></tr>
<tr><td colspan="2">NUMBER OF POSITIONS</td><td colspan="4"></td></tr>
<tr><td colspan="6"></td></tr>
<tr><td>CANDIDATE</td><td>Name</td><td colspan="4"></td></tr>
<tr><td></td><td>Contact details</td><td colspan="4"></td></tr>
<tr><td colspan="5">ESTABLISH THE NEED/URGENCY</td><td>☐</td></tr>
<tr><td colspan="6"></td></tr>
<tr><td colspan="6">JOB AND PERSON SPECIFICATION</td></tr>
<tr><td colspan="2">Is there a formal job specification?</td><td>Yes</td><td>☐</td><td>No</td><td>☐</td></tr>
<tr><td colspan="5">ROLE DESCRIPTION</td><td>☐</td></tr>
<tr><td colspan="6"></td></tr>
<tr><td colspan="5">ESSENTIAL</td><td>☐</td></tr>
<tr><td colspan="6"></td></tr>
<tr><td colspan="5">DESIRED</td><td>☐</td></tr>
<tr><td colspan="6"></td></tr>
<tr><td colspan="5">HIRING PROCESS</td><td>☐</td></tr>
</table>

What is the process that will happen from this conversation to the employee's starting day at work?				
SALARY/PAY RATE				☐
FULL JOB SPECIFICATION	Yes	☐	No	☐
INTERVIEW SLOTS	Yes	☐	No	☐
START DATE (PROVISIONAL)	Yes	☐	No	☐
BUDGET APPROVED	Yes	☐	No	☐
Agree exact time/date of next contact and actions for each party.				

Annex II

MANAGING TRAINING PROGRAMMES

This annex provides examples of factors that can assist with the management of training programmes in nuclear organizations.

II–1. FACTORS TO ASSIST LINE MANAGERS WITH THE OWNERSHIP AND CONTROL OF TRAINING

The following factors may help to enhance line management ownership and control of training:

— A written training policy document needs to be endorsed and communicated by senior managers to all nuclear facility staff. Written training procedures need to be available and used.
— 'Own' the training programmes for their personnel. The maintenance manager must feel responsible for the maintenance of the training programme, the operations manager for the operations training programme, etc. These managers need to be as responsible for making provision for training programmes as they are for providing resources for other needs of their organization.
— Meet with training staff to communicate and discuss training needs, define the scope and content of new training courses and review the appropriateness of existing courses.
— Review systematic approach to training (SAT) analysis phase data and all training materials during the SAT design and development phases and provide written comments on the materials. The training group needs to respond in writing to these comments to the satisfaction of the line manager.
— Provide on-the-job training and assessment of employee competence to perform tasks. Supervisors in the line organization need to ensure the trainees' completion of the training programme.
— Be held accountable for the adequacy of training and performance of their personnel. If errors occur due to insufficient competence, the line supervisor needs to analyse and understand the reasons for these occurrences; suggestions need to be made, in conjunction with the training staff, for suitable preventive and corrective actions, and the necessary actions implemented.

- Be able to demonstrate that both initial and continuing training are satisfying the performance requirements for the training programmes and the organization's goals.
- Support the use of peer reviews and self-assessment processes to ensure their training programmes are robust and are meeting the standards set in the training policy and training procedures.
- Be responsible for identifying the necessary training requirements to support improving performance. This will be achieved through the formal performance management process and/or through liaison with the training organization and through the relevant training committees.
- Be aware of the employee's personal development plan and also the organizational links to succession planning and workforce planning. This is particularly important when considering the long lead times associated with these two important activities.
- A network of training coordinators who facilitate interactions between the facility departments and training organization needs be considered.
- Training organization personnel need to meet with line managers, at least annually, in formal meetings with clear objectives to evaluate the training programmes, schedule training for the next year and agree on long term training strategies.
- The plant manager, operations manager or other senior plant supervisors need to periodically observe classroom and practical training, and assessments of competencies, particularly simulator based or on-the-job training examinations.
- The training manager needs to meet regularly with the plant manager to receive advice and direction. Supervisors in the training group need to meet regularly with their peers from plant and nuclear support groups, at nuclear facility training review committees, for example, to discuss training programmes.
- Line organization personnel need to participate in training development from the analysis phase through to the evaluation phase of the SAT process, and also in the development and revision of training procedures.

Annex III

LIST OF CONSIDERATIONS TO ENSURE THAT PAY AWARDS CAN OPERATE IN A TIMELY AND EFFECTIVE MANNER

This annex provides considerations to ensure that pay awards operate in a timely and effective manner.

III–1. CONSIDERATIONS

Procedures and processes need to be developed to ensure that pay awards can operate in a timely and effective manner. The following are a list of considerations to facilitate the process:

- Ensure the pay review process is treated as one aspect of the overall employee 'value proposition', and firmly linked with other human resources (HR) processes such as culture and values, performance management and talent and succession planning.
- Make sure planning commences well in advance of pay review implementation dates, with a specific emphasis on effective liaison between the reward/HR function and finance/the business at all times, as the latter are likely to have primary responsibility for employee costs.
- Collating and interpreting data on issues such as inflationary pressures and market pay rates at both a local and global level is a critical part of the planning process and can be difficult and time-consuming. Ensure access to better or more tailored data, in addition to drawing on published sources such as government data or privately produced salary surveys.
- While care and diligence needs to be taken in matching the salary review process as closely as possible with business needs, individual aspects of the process need to be kept as clear and simple as possible.
- The use of appropriate management tools and technical backup helps to support the decision making process.
- Following each review, discussions are concluded, and conclusions made on learning the lessons of that review, including the implications of any changes required by corporate governance and regulators together with identified design changes.
- There is clarity in the definitions of roles and responsibilities; who proposes, endorses and approves awards and the respective roles of line managers, remuneration committees, finance and HR.

- There is clear advance planning of the salary review process for the year ahead for the reward function’s workload.
- Salary review arrangements support broader reward management changes and are used as a vehicle for particular reward initiatives or to capitalize on wider benefits such as the opportunity to reinforce employer branding when communicating the pay award.
- The time and efforts expended on the salary review process are well spent, given that the effective undertaking is deemed essential to the success and well-being of the organization and its employees.

Annex IV

MANAGING PERFORMANCE

This annex provides considerations for managing performance at the organizational, process, job and individual levels.

IV–1. ORGANIZATIONAL LEVEL ATTRIBUTES FOR ACHIEVING RELIABLE PERFORMANCE

- Senior managers establish expectations for excellence in performance, safety culture and intolerance for process and workplace deficiencies. Strategic and business plans define goals, objectives, resources, performance measurements and supervision.
- Teamwork and healthy relationships result from a culture of trust, respect and fairness.
- Communication of information that can affect performance is highly valued.
- Managers routinely communicate and reinforce desired values and behaviours through observation, coaching, counselling, rewards and performance feedback. These managers also seek frequent input for performance improvement as well as trend performance.
- A process based management system is established at the nuclear facility, and self-assessment, a corrective action programme and management of human resources are integrated into the management system.

IV–2. PROCESS LEVEL ATTRIBUTES FOR ACHIEVING RELIABLE PERFORMANCE

- Processes are implemented as designed and are periodically assessed to eliminate weaknesses that could affect performance. Processes that are important for reliable performance include planning and scheduling, clearance tagging, configuration management, work control and use of operating experience.
- Work preparation and pre-job briefings identify critical actions and specific error prevention tools; potential human errors and their effects on the facility; contingency plans; and applicable operating experience.
- Procedures and other work documents are verified and validated for accuracy and usability. Deficiencies are corrected promptly.

— Changes in work plans and work schedules are critically reviewed for conditions that can provoke error or allow an undesirable effect on the facility.
— Initial and continuing training provides knowledge of error-prevention techniques, an understanding of their bases and importance, and reinforces the value of defences; this includes opportunities to practise using specific error prevention tools.
— Effective, corrective action programme (which might not necessarily involve root cause analysis) is used for improving processes.
— Root cause analysis, supplemented by performance improvement investigations, identifies the organizational, process and individual contributors to events.

IV–3. JOB LEVEL CONDITIONS THAT CONTRIBUTE TO RELIABLE PERFORMANCE

— Goals, roles and responsibilities for the assigned task are discussed and understood before work begins.
— Assigned personnel are technically qualified for the task, and are physically and mentally ready to perform the work.
— Job site conditions are properly established to enable qualified personnel to accomplish work assignments successfully.
— Job site conditions and potential consequences are carefully evaluated to reinforce desired work behaviours, to reduce the potential for human error.
— Work preparation and pre-job briefings are conducted commensurate with the risk of the work activity.
— A variety of defence in depth measures are used at the job site, commensurate with the risk of the work activity, to reduce the probability of error, as well as to mitigate the effects of and provide for recovery from error.
— Critical steps within the task are identified and the specific error prevention tools to be used to preclude an event of consequence are discussed.

IV–4. INDIVIDUAL VALUES AND BEHAVIOURS NEEDED FOR ACHIEVING RELIABLE PERFORMANCE

— Individuals maintain situational awareness, they are watchful for conditions or activities that can have an undesirable effect on facility performance or personnel safety, and they do not proceed if faced with uncertainty.

— Procedures and other work documents are used as intended; error-prevention techniques are understood, and appropriately and rigorously applied to task specific situations.
— Deficiencies and suggested improvements in processes, documents, equipment and the workplace are reported promptly.

ABBREVIATIONS

CSR	corporate and social responsibilities
HR	human resources
HRM	human resource management
KSAs	knowledge, skills and attitudes
PDPs	personal development plans
R&D	research and development
SAT	systematic approach to training
SMART	specific, measurable, achievable, realistic and timely
STEM	science, technology, engineering and mathematics

CONTRIBUTORS TO DRAFTING AND REVIEW

Bastos, J.	International Atomic Energy Agency
Bucholz, M.	Gesellschaft für Anlagen und Reaktorsicherheit, Germany
Dieguez Porras, P.	International Atomic Energy Agency
Drury, D.	International Atomic Energy Agency
Earp, S.	EDF Energy, United Kingdom
Fanjas, Y.	International Institute of Nuclear Energy, France
Françoise, C.	Canadian Nuclear Safety Commission, Canada
Franyo, I.	Paks Nuclear Power Plant, Hungary
Halt, L.	International Atomic Energy Agency
Islam, M.	Bangladesh Atomic Energy Commission, Bangladesh
Isotalo, J.	Teollisuuden Voima Oyj, Finland
Ivanytskyy, V.	National Nuclear Energy Generating Company, Ukraine
Jubin, J.-R.	International Atomic Energy Agency
Karezin, V.	ROSATOM, Russian Federation
Kern, K.	International Atomic Energy Agency
Kroshilin, A.	VNIIAES, Russian Federation
Kumar, S.	Kaiga Atomic Power Station, India
Mallam, S.	Nigeria Atomic Energy Commission, Nigeria
Mathews, L.	EDF Energy, United Kingdom
Molloy, B.	Consultant, Ireland
Mortin, S.	Consultant, United Kingdom
Mougel, B.	EDF, France

Nagibina, E.	ROSATOM, Russian Federation
Nam, Y.	Korea Atomic Energy Research Institute, Republic of Korea
O'Sullivan, P.	International Atomic Energy Agency
Palmer, D.	Consultant, United Kingdom
Pilatkowski, J.	Ministry of Energy, Poland
Skorvagova, I.	Nuclear Regulatory Authority, Slovakia
Szabo, V.	Nuclear Regulatory Authority, Slovakia
Thomas, C.	Thomas Thor, Netherlands
Tiron, C.	Nuclearelectrica, Romania
Van Sickle, M.	International Atomic Energy Agency
Widheden, B.	KSU Ringhals Nuclear Power Plant, Sweden
Wijewardane, S.	International Atomic Energy Agency
Zhang, Z.	Harbin Engineering University, China
Zvar, M.	Krško Nuclear Power Plant, Slovenia

Technical Working Group Meeting

Vienna, Austria: 25–27 September 2018

Consultants Meetings

Vienna, Austria: 6–8 February 2018; 11–15 February 2020

Structure of the IAEA Nuclear Energy Series*

Nuclear Energy Basic Principles
NE-BP

Nuclear Energy General Objectives
NG-O

1. Management Systems
NG-G-1.#
NG-T-1.#
2. Human Resources
NG-G-2.#
NG-T-2.#
3. Nuclear Infrastructure and Planning
NG-G-3.#
NG-T-3.#
4. Economics and Energy System Analysis
NG-G-4.#
NG-T-4.#
5. Stakeholder Involvement
NG-G-5.#
NG-T-5.#
6. Knowledge Management
NG-G-6.#
NG-T-6.#

Nuclear Reactor Objectives**
NR-O

1. Technology Development
NR-G-1.#
NR-T-1.#
2. Design, Construction and Commissioning of Nuclear Power Plants
NR-G-2.#
NR-T-2.#
3. Operation of Nuclear Power Plants
NR-G-3.#
NR-T-3.#
4. Non Electrical Applications
NR-G-4.#
NR-T-4.#
5. Research Reactors
NR-G-5.#
NR-T-5.#

Nuclear Fuel Cycle Objectives
NF-O

1. Exploration and Production of Raw Materials for Nuclear Energy
NF-G-1.#
NF-T-1.#
2. Fuel Engineering and Performance
NF-G-2.#
NF-T-2.#
3. Spent Fuel Management
NF-G-3.#
NF-T-3.#
4. Fuel Cycle Options
NF-G-4.#
NF-T-4.#
5. Nuclear Fuel Cycle Facilities
NF-G-5.#
NF-T-5.#

Radioactive Waste Management and Decommissioning Objectives
NW-O

1. Radioactive Waste Management
NW-G-1.#
NW-T-1.#
2. Decommissioning of Nuclear Facilities
NW-G-2.#
NW-T-2.#
3. Environmental Remediation
NW-G-3.#
NW-T-3.#

(*) as of 1 January 2020
(**) Formerly 'Nuclear Power' (NP)

Key

BP:	Basic Principles
O:	Objectives
G:	Guides and Methodologies
T:	Technical Reports
Nos 1–6:	Topic designations
#:	Guide or Report number

Examples

NG-G-3.1:	Nuclear Energy General (**NG**), Guides and Methodologies (**G**), Nuclear Infrastructure and Planning (topic **3**), **#1**
NR-T-5.4:	Nuclear Reactors (**NR**), Technical Report (**T**), Research Reactors (topic **5**), **#4**
NF-T-3.6:	Nuclear Fuel (**NF**), Technical Report (**T**), Spent Fuel Management (topic **3**), **#6**
NW-G-1.1:	Radioactive Waste Management and Decommissioning (**NW**), Guides and Methodologies (**G**), Radioactive Waste Management (topic **1**) **#1**

No. 26

ORDERING LOCALLY

IAEA priced publications may be purchased from the sources listed below or from major local booksellers.

Orders for unpriced publications should be made directly to the IAEA. The contact details are given at the end of this list.

NORTH AMERICA

Bernan / Rowman & Littlefield
15250 NBN Way, Blue Ridge Summit, PA 17214, USA
Telephone: +1 800 462 6420 • Fax: +1 800 338 4550
Email: orders@rowman.com • Web site: www.rowman.com/bernan

REST OF WORLD

Please contact your preferred local supplier, or our lead distributor:

Eurospan Group
Gray's Inn House
127 Clerkenwell Road
London EC1R 5DB
United Kingdom

Trade orders and enquiries:
Telephone: +44 (0)176 760 4972 • Fax: +44 (0)176 760 1640
Email: eurospan@turpin-distribution.com

Individual orders:
www.eurospanbookstore.com/iaea

For further information:
Telephone: +44 (0)207 240 0856 • Fax: +44 (0)207 379 0609
Email: info@eurospangroup.com • Web site: www.eurospangroup.com

Orders for both priced and unpriced publications may be addressed directly to:

Marketing and Sales Unit
International Atomic Energy Agency
Vienna International Centre, PO Box 100, 1400 Vienna, Austria
Telephone: +43 1 2600 22529 or 22530 • Fax: +43 1 26007 22529
Email: sales.publications@iaea.org • Web site: www.iaea.org/publications